WITHDRAWN

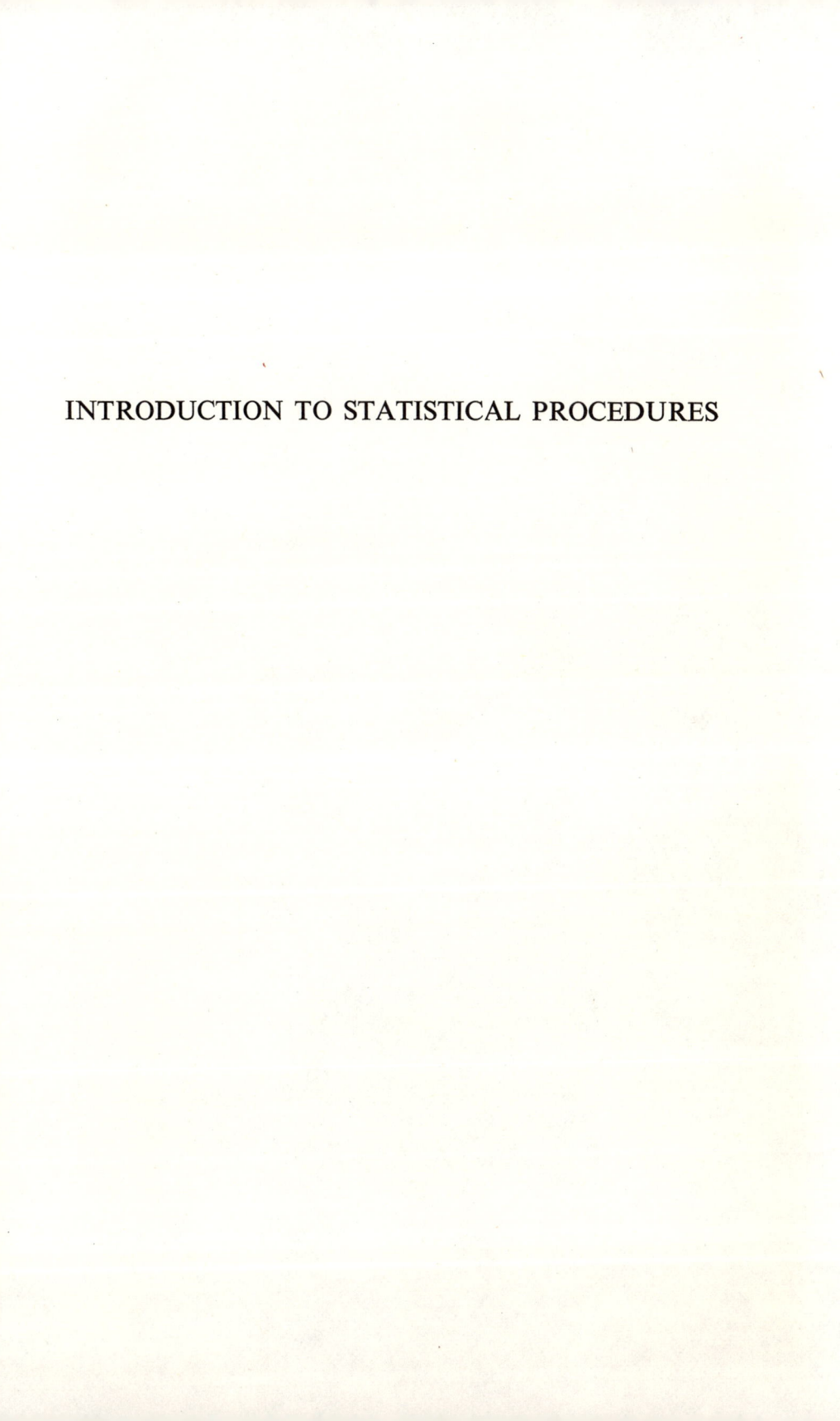

INTRODUCTION TO STATISTICAL PROCEDURES

Introduction to Statistical Procedures:
with Computer Exercises

PAUL R. LOHNES
State University of New York at Buffalo

WILLIAM W. COOLEY
University of Pittsburgh

John Wiley & Sons, Inc.
NEW YORK   LONDON   SYDNEY   TORONTO

Copyright © 1968 by John Wiley & Sons, Inc.

All rights reserved. No part of this book may be reproduced by any means, nor transmitted, nor translated into a machine language without the written permission of the publisher.

Library of Congress Catalog Card Number: 68-24797
SBN 471 54348 9
Printed in the United States of America

to Professor Phillip J. Rulon

of Harvard University

who made statistics a royal adventure for his students

Preface

This is a new approach to a one-semester course in introductory statistics for undergraduates and graduate students in the behavioral sciences and education. Perhaps the most unusual feature of the course is that it teaches the student how to read and use computer programs. A less unusual but thoroughly important feature is the emphasis placed on basic statistical theory, as opposed to statistical methods. Yet another feature is the presentation of selected examples of behavioral research and "real live" data for student practice.

Although the computer revolution has been with us for some time, statistics books tend to ignore the implications of this technological breakthrough. We believe that there is no excuse for burdening statistics students with arithmetic chores that are now properly the business of artificial intelligence. In this text we teach the student how to turn the data analysis of statistics over to a computer in order to free his intelligence for the study of human problems of theory, design, and interpretation. The student also acquires the rudiments of a programming skill which is indispensable today in advanced statistics. Perhaps this alone will encourage some students to continue beyond the first course.

The main emphasis of the book is on statistics as an epistemology rather than as a collection of methods. The student is introduced to a challenging theory of knowledge. Its basic notions of induction, sampling, randomizing, estimating, and inferring, and its basic theorems—the law of large numbers and the central-limits theorem—are explained and illustrated. As an innovation in the teaching of applied statistics, a series of computer laboratory experiments applying Monte Carlo simulation techniques provides empirical demonstrations of basic principles and distributions. These experiments permit the student to develop important insights into statistics that are difficult to obtain with conventional teaching methods. Throughout the text, but especially in the last chapter, every effort is made to persuade the student that he has begun an intellectual journey which he cannot afford to terminate after one semester.

Among the professors with whom we studied statistics, and from whom we gained our love of the subject, are John B. Carroll, Palmer O. Johnson, Phillip J. Rulon, William B. Schraeder, and David V. Tiedeman. We hope they will not be appalled by this by-product of their pedagogy. We thank John C. Flanagan for allowing us to borrow data and examples from Project TALENT. Many colleagues and students assisted in this undertaking with counsel and criticism, but Bary G. Wingersky, Silas Halperin, James E. Carlson, and Richard Ferguson were particularly helpful. We are also indebted to our wives, who patiently put up with absent husbands on weekends so that this work could be done. Without the urging of our editor, Joseph F. Jordon, we would not have written this text. Arthur P. Coladarci and Ellis B. Page provided a thorough and thoughtful editing for the entire text.

Finally, we thank Freda Womer, Violet Marcus, Janet Combs, and Linda Schirm for their labors in manuscript preparation at various stages of development.

To all these we express our gratitude.

January 1968

Paul R. Lohnes
William W. Cooley

Notes on Teaching

It is assumed that the instructor will provide about fifteen weekly lectures, explaining and augmenting the contents of the text. He should schedule a minimum of ten laboratory periods, to be devoted to the computer experiments. On campuses where there is a small, open-shop computer installation, such as an IBM 1620, an ideal arrangement is to take over the computer for an hour for each of these periods, so that students may be present while the computations are done. However, where the installation is a large, closed-shop one, such as an IBM 7090, the laboratory meetings may take place in a regular classroom, with the instructor or an assistant collecting computer input (usually a few setup cards) from students, returning computer output listings to students, and discussing the computer programs, the outputs, and the writing assignments based on the outputs. In a closed-shop environment it is very desirable to schedule the first laboratory period at the computer center, incorporating a guided tour and an on-line submission of a modest program (for example, List 4.2). Students will require access to a card punch during the term, although the time requirement will be slight. The programs have been arranged to print reports, copies of which can be made for all the students.

A preliminary edition of this text was used at the University of Pittsburgh with both an IBM 7090 and the new time-sharing IBM 360/50. In the case of the latter, each student did his own laboratory exercises at a remote terminal, where the text programs were modified into a more self-instructional mode. This approach is highly recommended if such computer equipment is available. Persons interested in these later developments should contact W. W. Cooley, School of Education, University of Pittsburgh, 15213.

The instructor with computer experience will recognize that the FORTRAN programs in this book are not complicated all-purpose production programs; they are kept simple for instructional purposes. You will probably have only one question about them, "Where can I get program decks?" Program decks may be purchased by users desiring to avoid a local punching chore by contacting P. R. Lohnes at the Department of Educational Psychology, S.U.N.Y.A.B., Buffalo, N.Y. 14214.

The instructor who will be breaking in on the computer with these programs may be assured that it is simplicity itself to get these programs running. As far as we know, almost every college computer is capable of compiling FORTRAN, even if it is not an IBM machine. The programs, as presented here, may require some slight modifications, particularly and perhaps only in the input-output statements, to meet the conventions of your local computer center. One session with a consultant in your computer center will suffice to locate and correct these difficulties. It is really very easy to become a computer user, as we hope this text will prove to many, many students of statistics.

We urge the instructor to enforce a productive computing-thinking ratio by requiring written interpretations of the experiments. Listings of outputs from fresh replications of some of the experiments can be distributed during examination periods to provide the material for challenging test questions.

FORTRAN programs LIST 4.3 and LIST 9.3 are the real workhorse routines in this program set, because they prepare grouped-data frequency distributions and chi-square goodness of fit tests from the results of the Monte Carlo generation programs.

Contents

Computer Programs

Chapter One Statistical Procedures in Research
1.1 Educational research — 1
1.2 An educational experiment — 2
1.3 Project TALENT, a sampling survey — 7
1.4 The punched card — 14
1.5 Summary — 16

Chapter Two Describing Univariate Distributions
2.1 Sources and types of data — 18
2.2 The frequency distribution — 21
2.3 Central tendency — 27
2.4 Statistics and parameters — 28
2.5 Summary — 31

Chapter Three Variability in Measurements
3.1 Variability — 33
3.2 Variance — 35
3.3 Standard deviation — 39
3.4 Standard scores — 43
3.5 Summary — 47

Chapter Four Computing Statistical Analyses
4.1 Introduction to digital computer — 49
4.2 The FORTRAN language — 54
4.3 Computing distribution statistics — 68
4.4 Summary — 72

Chapter Five Computer Experiments with Random Numbers
5.1 Random number generation — 79
5.2 A Monte Carlo study of the variance estimates — 87
5.3 Summary — 89

Chapter Six Theoretical Distributions
- 6.1 An experiment — 93
- 6.2 The binomial distribution — 96
- 6.3 The role of theoretical distributions in statistics — 100
- 6.4 Summary — 104

Chapter Seven Normal Curve Theory
- 7.1 The normal distribution — 107
- 7.2 The central-limit theorem — 116
- 7.3 Departure from normality — 122
- 7.4 Summary — 123

Chapter Eight Standard Errors and Statistical Inference
- 8.1 Standard error of the mean — 126
- 8.2 Testing a hypothesis about a mean — 132
- 8.3 Significance of difference between two means — 138
- 8.4 Summary — 140

Chapter Nine Goodness of Fit
- 9.1 The chi-square distribution — 144
- 9.2 The goodness of fit test — 149
- 9.3 Applications to Monte Carlo results — 153
- 9.4 Summary — 157

Chapter Ten Other Uses of Chi-Square Distribution
- 10.1 Testing a hypothesis about a variance — 161
- 10.2 The significance of a difference between proportions — 162
- 10.3 Types of errors — 167
- 10.4 Summary — 171

Chapter Eleven Small Sample Statistics
- 11.1 The t distribution — 174
- 11.2 Confidence intervals — 177
- 11.3 Comparison of two means — 177
- 11.4 Monte Carlo demonstrations of t distribution — 179
- 11.5 Application of t-test — 181
- 11.6 Summary — 182

Chapter Twelve Exact Randomization Tests
- 12.1 The randomization outcomes of an experiment — 186
- 12.2 Summary — 194

Chapter Thirteen Statistical Prediction
- 13.1 The prediction problem — 198
- 13.2 Prediction through regression — 204
- 13.3 The standard error of estimate — 209
- 13.4 Example using Project TALENT data — 213
- 13.5 Summary — 213

Chapter Fourteen Correlation
- 14.1 A measure of relationship — 215
- 14.2 The correlation coefficient — 217
- 14.3 The bivariate normal distribution — 219
- 14.4 Other interpretations of the correlation coefficient — 221
- 14.5 Tests of inference involving r — 222
- 14.6 The correlation matrix — 224
- 14.7 Summary — 225

Chapter Fifteen An Overview of Statistics
- 15.1 Inferring and estimating relationships — 230
- 15.2 Random sampling and random assignment — 237

References — 240

Appendix
- A Symbol Conventions — 243
- B Project TALENT Data — 245
- C Answers to Selected Exercises — 259
- D Summary of Mathematical Concepts Necessary for Understanding Elementary Statistical Concepts — 267
- E Theoretical Distributions — 274

Index — 277

Computer Programs

LIST	4.1	Transformed Scores	74
	4.2	Univariate Distribution	75
	4.3	Subroutine Unidis	77
	5.1	Function Random(K)	90
	5.2	Uniformity of Random Fractions	91
	5.3	Monte Carlo on Sample Variance	92
	6.1	Binomial Distribution	106
	7.1	Function Randev(K)	125
	7.2	Test Function Randev(K)	125
	8.1	Monte Carlo on Sample Mean	143
	9.1	Chi-Square Generator	158
	9.2	Goodness of Fit	160
	9.3	Subroutine Goofit	160
	11.1	Monte Carlo on T Test	184
	11.2	T Test	185
	12.1	Exact Randomization Test	195
	14.1	Monte Carlo on Correlation	227
	14.2	Correlation and Regression	228

INTRODUCTION TO STATISTICAL PROCEDURES

CHAPTER ONE

Statistical Procedures in Research

1.1 Educational research

The approach to learning statistical procedures used in this book will be useful throughout the behavioral sciences. The examples, however, have been drawn from the field of educational research. One reason for this particular application is that both authors are active educational researchers and are familiar with that field. Furthermore, and more importantly, educational research is an especially suitable source of examples for an introduction to statistical methods because of the multidisciplinary nature of research in education. An investigator studying students, teachers, and schools is sometimes a sociologist, frequently a psychologist, and at times anthropologist, psychiatrist, historian, economist, or architect.

The recent explosive expansion of activity in educational research, girded strongly by Federal and private financial support, is another justification for our attention to this field. This increased activity has brought with it many problems. One of chief concern is that there is now more money available to conduct educational research than there are people prepared to design, execute, report, and utilize such investigations. This need has not been met despite the fact that the new support has attracted to educational problems many competent scholars from many disciplines who previously had neither the money nor inclination to devote themselves to the study of educational problems.

Increased statistical sophistication will help reduce another difficulty in current educational research. The fact is that rather static models for guiding inquiry have dominated research activity in education. Educational experiments up to 1960 largely were efforts to evaluate two quite gross teaching strategies in two class sections of a course by comparing the test scores of the two groups at the end of the course. This model merely allows the investigator to say that Method A is better than B, B is better than A, or there is no difference between A and B. Such experiments have had little impact on educational practice for several reasons. One reason is that they have been of such limited scope that we quickly can find sound reasons for discounting the generalizability of the results to other situations. Furthermore, the results have not been convincing because of the primitive nature of the test used as a criterion measure, and the fact that the test selected frequently had little demonstrated relevance to the objectives of the instruction. Perhaps a more serious shortcoming of the classical "A versus B" model is that the only product is the gross comparison of the average work of the two groups being compared. This leaves the investigator with little if any insights regarding how either Method A *or* B might be improved as a result of his investigation.

Fortunately, the picture of educational research today is improving. Many exciting experiments and surveys are being conducted in which much can be learned about the vast enterprise of education. The student who has prepared himself to research educational problems can now look forward to abundant opportunities for support of his work and a growing body of highly competent colleagues.

This book is intended to begin the training of the student entering this dynamic field. It will also be useful in preparing educators who do not expect to conduct research but who need to evaluate and utilize the products of those who do.

Statistical analysis plays a dominant role in two major types of educational research activities: *surveys* and *experiments*. Although the student may not now clearly see the distinction between them, it should be quite clear by the time he has completed this course. He will also eventually see that there are many varieties of surveys and experiments. Some surveys, for example, involve simple counts on a single *variable* (for example, class size in American high schools), while others comprise measures of many variables (for example, class size, cost per pupil, and student achievement scores) and their interrelationships. To introduce the major distinction between surveys and experiments we have selected and will describe one of each. One of the two examples has been chosen also because it provides appropriate data for student practice. That is, in the following two sections of this chapter two studies are described in some detail and data are provided which were obtained in one of them. These data will serve as practice data so that the student can "play scientist" as he begins to master his new statistical tools.

1.2 An educational experiment

An educational experiment occurs when a scientist assigns subjects at random to two or more treatment groups, imposes a different educational treatment on each group, and then measures the groups to see whether the different treatments have produced important differences among the groups on one or more outcome variables.

Random assignment means that each subject has as much chance of being assigned to any one treatment as to any other treatment. One reason the scientist has for making random assignments of subjects to treatment groups is to eliminate systematic bias from the vast number of variables on which subjects may differ. If he does not do so, that bias at the beginning of the experiment will probably influence the outcome measurements. Random sampling and a method of generating random assignments with the aid of a computer are discussed in Chapter 5.

When it comes to the selection of different educational procedures to be tested against each other for relative efficiency in an experiment, educational researchers suffer from an embarrassment of riches. Teachers in any given content field for any given age level have access to such a variety of teaching methods and teaching materials that it is difficult to decide what innovations to evaluate relative to what standard prescriptions. This text is going to call to your attention some of the technical features of good experimental designs, but good design does not guarantee a worthwhile experiment. The scientist has to have the wisdom to phrase important research questions as well as the ability to plan experiments that can answer his questions. In educational scholarship it is difficult to know what the most important questions are that lend themselves to research. We suggest that you scan a recent volume of an educational research journal (perhaps *Journal of Educational Psychology* or *American Educational Research Journal*) and list the research questions covered by the articles. Does your list contain some seemingly important problems? Does anything on the list seem rather trivial? While you are about this task, figure out what percentage of the articles employ statistics in the analysis of research data. You ought to find ample evidence that you need to learn some statistics if you plan to read educational research reports.

After the period in which the groups of subjects have received different educational treatments, the scientist measures all subjects on one or more outcome variables (frequently these are achievement tests), and analyzes the scores to decide whether the groups differ in outcomes. We can best define *statistics* by considering the problems the scientist faces in performing this analysis of outcome measurements. First, he has to characterize each group by some summary of the scores for the subjects in the group. Computing an average score for the group is an example of such a summary. Statistics is in part the science of summarizing masses of measurements to obtain useful summary *descriptions of trends in the data*, such as characterizations of groups and of differences among groups. Second, the scientist has to decide whether the differences among the groups on the descriptive statistics could be due simply to the failure of his initial randomization procedure to equate the groups prior to the different treatments. He knows that if he had given precisely the same treatment to all the groups, they would nevertheless differ in outcomes by chance alone, as a result of accidents in his randomizing procedure. The other part of statistics is the science of making *inferences about the significance of trends in the data*, such as observed differences among groups. In short, statistics is the science of making *descriptions* and *inferences* from measurements.

One result from your study of introductory statistics will be that you will come to appreciate that randomization always plays some role in the collection of data for statistical analyses. You will see that the scientist buys the power of statistical analysis with the currency of randomization.

We are going to give you a brief description of an educational experiment called Project HOSC (History of Science Cases).[1] We ask you to study it carefully to learn what you can about the experiment and, specifically, to locate the special concepts of research design that are unknown to you. These concepts are a small sample of the everyday stock in trade of working educational researchers.

The HOSC Instruction Project was an experimental inquiry into the high school science curriculum. One of the two investigators, L. E. Klopfer, had developed a set of historical case studies on topics in the histories of biology, chemistry, and physics. The investigators contended that conventional high school science courses were preoccupied with the results of science as systematic bodies of knowledge and failed to convey adequate understandings of science as a process and an institution, or of scientists as people with values, interests, and other personality characteristics. By focusing on biographical materials organized around the histories of major discoveries in science, the HOSC units were intended to place appropriate emphasis on these neglected curriculum objectives.

The research hypothesis of the HOSC Project was that high school science classes that studied one or more HOSC units would understand more about science and scientists than classes that did not study HOSC units. This hypothesis was based on the assumption that students are more likely to acquire learnings that the teacher deliberately plans than they are to acquire the same learnings incidentally.

The investigators recognized that science teachers differ in their own understandings of science and scientists and, from the position that students learn more from teachers who know more, they proposed a second hypothesis: high school science classes that studied with teachers who were above average in understanding of science would learn more understanding of science than classes that studied with teachers who were below average.

A third hypothesis which the investigators wished to test was that the amount of learning of understanding of science would depend also on which science subject (biology, chemistry, or physics) was being taught to the class.

The HOSC units were to be introduced into some classes at the expense of interrupting the normal flow of conventional instruction designed to produce

---

[1] For further details, see Klopfer and Cooley (1963). (References follow Chapter 15.)

knowledge of the results of science. Therefore the same reasoning that led to the first hypothesis also led to the fourth, which asserted that classes that did not study HOSC units would gain more content knowledge of science than would classes that studied HOSC units.

Here, then, is an intention to test four separate hypotheses simultaneously. You might ask whether it would be better for an experiment to test a single hypothesis. In a laboratory setting it is possible to keep experiments simple. In actual classroom situations things cannot be simple, and experiments sited in classrooms have to respect and capitalize on the complexity that exists.

A major problem for these investigators, as for all educational researchers, was to find ways of measuring the learning variables involved in their hypotheses. Measurement instruments selected for use in an experiment must have *validity*, which means that they must measure what they purport to measure, and not other things. They must also have *reliability*, which means that they must measure what they measure with sufficient precision. In reading any research report it is important to examine the evidence that the tests and other instruments used actually measure the variables the hypotheses deal with.

The investigators in the HOSC Project were able to select from among standard, published tests some of the instruments they needed. From Educational Testing Service they obtained the Cooperative Physics, Chemistry, and Biology Tests, which are measures of the conventional science knowledges expected as outcomes of the corresponding high school science courses. From World Book Company they obtained the Otis Mental Ability Test, which is a widely used test of general academic aptitude. But when it came to the two problems of measuring in teachers and pupils the kinds of understandings of science and scientists the HOSC units were designed to produce, it was necessary for the investigators to develop their own instruments. The development of a standardized test involves a lengthy series of highly technical (and statistical) activities which the specialized training of the educational or psychological researcher prepares him for. Suffice it to say that the investigators emerged with teacher and student versions of tests of the required traits. The test for students has been made available by Educational Testing Service and has figured in several other science education researches. It is called TOUS, for *Test On Understanding Science*.

When the hypotheses to be tested by an experiment are specified, and the instruments for measuring the variables involved in the hypotheses are ready, the next problem is the *design* of the experiment. The design is the plan for grouping subjects for experimental treatment. In the HOSC Project, classroom groups of science students were to be the units of analysis. The design called for separating the available classes into three groups, according

Table 1.1

Science Subject	Teacher's Understanding of Science		Row Totals
	High	Low	
Biology	HOSC Method [10 classes] — Non-HOSC [10 classes] (Randomized)	HOSC Method [10 classes] — Non-HOSC [10 classes] (Randomized)	40 classes
Chemistry	HOSC Method [10 classes] — Non-HOSC [10 classes] (Randomized)	HOSC Method [10 classes] — Non-HOSC [10 classes] (Randomized)	40 classes
Physics	HOSC Method [7 classes] — Non-HOSC [7 classes] (Randomized)	HOSC Method [7 classes] — Non-HOSC [7 classes] (Randomized)	28 classes
Column totals	54 Classes	54 Classes	108 classes

to subject matter (biology, physics, and chemistry). Each of these groups was divided into two, according to whether the teacher's tested understanding of science made her a "high" or a "low" teacher. Finally, in each of these six groups of classes, half was randomly assigned to receive the HOSC enrichment treatment, and the other half was assigned to receive the conventional curriculum treatment only. The resulting design contained twelve cells, reflecting random assignment of classes to treatments from within teacher–subject combinations. Table 1.1 illustrates the design. Every class was tested with the Otis Mental Ability Test and TOUS in the first month of the school year (called *pretests*). The science instruction followed the normal course until March for the non-HOSC classes, but somewhere in the interim the HOSC classes were taught two HOSC units. In March all classes were tested again with TOUS and the appropriate COOP Science Test (called *posttests*).

The science classes that participated in the HOSC Project were from high schools across the United States. The investigators give the following description of their sample.

These 108 participating classes were widely distributed geographically. They included representatives from public and private schools, from small, medium-sized, and large high schools, from urban, suburban, and rural communities, and were taught by teachers of both sexes and with a broad variation of teaching experience. In general, the sample was representative of high school biology, chemistry, and physics classes in the United States today. (Klopfer and Cooley, 1963, p. 35.)

Notice that all classes were tested twice, *before* and *after* instruction by the different treatment methods. The reason for the pretests was that the investigators decided to take into account possible differences among the classes in initial abilities. This is a complicated idea, and an even more complicated statistical procedure. What the investigators did was "adjust" all the posttreatment class differences to remove any advantage or disadvantage a class might have as found in its pretest scores. The statistical method of making such adjustments in experimental outcomes on the basis of pretest scores is called "analysis of covariance." You will not learn this method until your second or third course in educational statistics, but by the end of this course you will know its basis, which is the "correlation" between pretest and posttest.

Notice that this mysterious "covariance" method of adjusting for class differences on one important input variable did not excuse the scientists from the requirement that they make random assignments of classes to HOSC and non-HOSC teaching methods. The legitimacy of all their statistical methods depends on the element of randomization.

The HOSC Project showed that the History of Science Cases made a difference in learning outcomes. On the average, the HOSC classes emerged with a superior understanding of science over the non-HOSC classes, as measured by the TOUS test. The experimental design also allowed the investigators to explore whether the effectiveness of the case histories depended in part upon certain conditions, such as type of science course or teacher or some combination of both. A more detailed analysis of the results also showed special strengths and weaknesses of the HOSC units. That is, some aspects of understanding science were better learned than other aspects. Educators require considerable training in measurement and statistics if they wish either to design or to read and evaluate such experiments.

1.3 Project TALENT, a sampling survey

In 1959 the U.S. Office of Education agreed to support a massive census of the abilities and other personality characteristics of American high school youth, and of the characteristics of their high schools. The project has been a joint effort of the American Institutes for Research and the University of Pittsburgh under the leadership of John C. Flanagan. A primary purpose of the study is to survey the talents of youth, in order to estimate the size of the manpower pools entering the various fields of work and to examine the personal characteristics of those pools (Flanagan et al., 1962a, p. 3). The relationships among the abilities, interests, and other characteristics of American youth, and the relations of these personality characteristics to aspects of the schools and the families of the youth are being determined. The project staff plan to keep in touch with the young people in the sample for at least twenty years in order that their life and career adjustment outcomes may become known and the degrees of predictability of these adjustments from the measurement variables of the original survey may be computed.

The scope of the undertaking is evidenced by the sample sizes involved: 440,000 youth in over 1300 high schools filled out more than two million answer sheets in two days of testing and inventorying. In addition, school officers supplied pages of information on each of the schools. The selection of the sample of the nation's high schools was planned to insure adequate representation of all types of schools, private and public, large and small, rural and urban, in all geographic areas. All students in grades nine through twelve in each of the participating schools were tested, yielding a sample of approximately 5 percent of the nation's high school youth in the spring of 1960. From the test and inventory results on this 5 percent sample it has been possible to make a very trustworthy estimate of corresponding parameters for the entire population of high school youth. This same type of sampling survey is employed by the Bureau of the Census for many of its detailed studies, and is the method of the public opinion polls, which figure so prominently in the news prior to national elections. In fact, the TALENT staff envision a permanent role for their organization in American educational research, with a new TALENT high school survey being taken every ten years, in the census year. Such a systematic, periodic inventory of youth and schools would permit the charting of trends in educational productivity in our society. Statistics finds a major application in the design and execution of such scientific sampling surveys.

The theory underlying Project TALENT rejects the notion that human talent is a single trait of general intelligence. It holds that "each person possesses his own unique pattern of potentials for learning to perform various types of activities important in our culture" (Flanagan et al., 1962a, p. 25). The theory also holds that a unique pattern of other personality attributes can be estimated for each student from inventories of interests, values, attitudes, and plans. Numerical scores are returned by all these measurement devices, and the unique pattern for a student is represented by a vector of scores (a row of numbers). The Project TALENT test battery, prepared especially for this survey, is outlined in Table 1.2. All the information about the students, their families, and their schools and communities collected by the project has been coded numerically and stored on magnetic

Table 1.2 Project TALENT Test Battery[a] (Aptitude, Achievement and Motive Tests)

Information Test

TALENT Code Number	Name	Number of Items
	PART I	
101	Screening	12
102	Vocabulary	21
103	Literature	24
104	Music	13
105	Social studies	24
106	Mathematics	23
107	Physical science	18
108	Biological science	11
109	Scientific attitude	10
110	Aeronautics and space	10
111	Electricity and electronics	20
112	Mechanics	19
113	Farming	12
114	Home economics	21
115	Sports	14
190	Part I total	(252)
	PART II	
131	Art	12
132	Law	9
133	Health	9
134	Engineering	6
135	Architecture	6
136	Journalism	3
137	Foreign travel	5
138	Military	7
139	Accounting, business, sales	10
140	Practical knowledge	4
141	Clerical	3
142	Bible	15
143	Colors	3
144	Etiquette	2
145	Hunting	5
146	Fishing	5
147	Outdoor activities (other)	9
148	Photography	3
149	Games (sedentary)	5
150	Theater and ballet	8
151	Foods	4
152	Miscellaneous	10
192	Part II total (inc. 10 misc. items)	(143)
	Parts I and II combined	(395)
100	Grand total	(395)
211	Memory for sentences	16
212	Memory for words	24
220	Disguised words	30
	English	(113)
231	Spelling	16

[a] Flanagan et al. (1962a), p. 223–25.

Table 1.2 (*Continued*)

Information Test

TALENT Code Number	Name	Number of Items
232	Capitalization	33
233	Punctuation	27
234	English usage	25
235	Effective expression	12
240	Word functions in sentences	24
250	Reading comprehension	48
260	Creativity	20
270	Mechanical reasoning	20
281	Visualization in two dimensions	24
282	Visualization in three dimensions	16
290	Abstract reasoning	15
	Mathematics	(54)
311	Part I. Arithmetic reasoning	16
312	Part II. Introductory high school mathematics	24
320	Subtotal (Parts I and II)	(40)
333	Part III. Advanced high school mathematics	14
334	High school subtotal (Parts II and III)	(38)
340	Total (Parts I, II, and III)	(54)
410	Arithmetic computation	72
420	Table reading	72
430	Clerical checking	74
440	Object inspection	40
600	*Student Activities Inventory*	(150)
601	Sociability	12
602	Social sensitivity	9
603	Impulsiveness	9
604	Vigor	7
605	Calmness	9
606	Tidiness	11
607	Culture	10
608	Leadership	5
609	Self-confidence	12
610	Mature personality	24
700	*Interest Inventory*	(205)
701	Physical science, engineering, mathematics	16
702	Biological science and medicine	8
703	Public service	11
704	Literary-linguistic	16
705	Social service	12
706	Artistic	7
707	Musical	5
708	Sports	8
709	Hunting and fishing	3
710	Business management	14
711	Sales	6
712	Computation	10
713	Office work	7
714	Mechanical-technical	15
715	Skilled trades	18
716	Farming	7
717	Labor	10
800	Miscellaneous: student information blank	(394)

tapes for computer processing. The adoption of numerical codes for information is a prerequisite for the application of most statistical methods in research.

Project TALENT is an example of a *research program*, staffed and equipped for a long-term assault on a succession of research questions. Some of the questions TALENT will answer in the future are not known today. A large amount has already been accomplished, as testified by the publication to date of several large volumes (Flanagan et al., 1962a; Flanagan et al., 1962b; Shaycoft et al., 1963; Flanagan et al., 1964; Flanagan et al., 1966; Lohnes, 1966).[2] The Project also has supplied research information for the Selective Service, the Peace Corps, the Job Corps, the National Merit Scholarship Corporation, and many other agencies. After you have completed this course, you will be in a position to read with understanding a variety of statistical findings conveyed in these reports.

In the HOSC experiment, the critical aspect of randomization was in the assignment of the teacher-class sections to one of the two treatment groups. The sample of science teachers available for the HOSC experiment did not constitute a random sample of all American high school science teachers. In such an experiment, the extent to which generalizations can be made from the results depends upon the extent to which relevant variables were directly included in the experiment itself. For example, if only biology teacher-class sections had been included, it would not have been possible to generalize these results to chemistry and physics teachers and students. However, since all three types were included as a factor in the design and analysis, it is possible to estimate whether the relative effectiveness of the HOSC units depends upon whether we are talking about biology, chemistry, or physics teacher-class sections.

In the HOSC experiment, two variables were of sufficient importance to include as stratification (classification) variables prior to the random assignment of class sections to treatments. These two stratification variables were science subject (biology, chemistry, or physics) and teacher's understanding of science (high or low). Inclusion of these two variables in the HOSC design allowed the experimenter to make generalizations about the HOSC treatment in terms of science subject matter and the teacher's understanding of science. Even if the HOSC experimenters had selected a completely random sample of American high school science teacher-class sections, generalizations regarding the effectiveness of the HOSC treatment would only be possible in terms of the factors included in the experimental design. Even if teachers from all parts of the country were included, we could not say that the HOSC treatment improved understanding of science regardless of region unless

---

[2] A complete list of TALENT publications can be obtained from the Project TALENT Office, Post Office Box 1113, Palo Alto, California 94302.

geographic region was actually included as a stratified factor in the design and analysis, as were type of science subject and teacher understanding. On the other hand, if teacher-class sections had not been randomly assigned to the HOSC treatment, no generalizations would have been possible regarding the effectiveness of the treatment.

Project TALENT does not involve any experimental treatments. In this survey the main objective is to estimate characteristics of the population of American high schools and the population of American high school students. In such a survey, the critical aspects of randomization involve the selection of the sample to be studied. One way of selecting such a sample might be to prepare a card for every high school in the United States, put the cards in a random fashion (such as shuffling a deck of cards in a game of poker), and then select every twentieth card. Such a procedure would yield a 5 percent sample of American high schools. If we then wanted to determine how many ninth-grade boys were planning to be physicists in the United States, we could count the number of ninth-grade boys with such plans in our 5 percent sample and multiply that result by 20 to estimate the total number in the United States with such plans.

One problem with such a sample is that there are many, many more small schools in this country than there are large schools.

The sample would be made up of students from more small schools than we would need in order to describe the characteristics of small schools in the country and too few large schools to be able to make confident generalizations about them. The survey designer might also want to make sure that he has an adequate sample of schools from all regions of the United States and that he has both schools with a large graduation rate (that is, high holding power) and schools with a low rate.

These were some of the considerations that the designers of Project TALENT faced when they decided how to sample American high schools for inclusion in Project TALENT. They decided to sort all of the public high schools by geographic region, holding power, and size (large, medium, and small). Then, to have an adequate sample of large schools, they selected one in thirteen large schools. So that they would not have too many small schools, they selected one in fifty such schools. Medium sized public high schools and private and parochial schools were selected in the ratio of one in twenty. This information is summarized in Table 1.3. In order to estimate the number of ninth-grade males planning to become physicists, the number in the large public schools is multiplied by 13, the number in the medium public schools by 20, the number in the small schools by 50, and the number in the private and parochial by 20. The sum of all these is then the estimate of the total number of ninth graders planning to become physicists in the entire nation.

Table 1.3 Sampling Design for Project TALENT

Public High Schools (Stratified by Geographic Region and Holding Power)	Sampling Ratio
Size	
Large	1 in 13
Medium	1 in 20
Small	1 in 50
Private and Parochial	1 in 20

It is important to note that although it is essential for randomization to be involved in the selection of the sample to be surveyed, the probability of being in the sample does not have to be the same for all members of the population. What is important is that every member has a known *nonzero* probability of being in the sample and that these different sampling ratios (that is, these different probabilities of being in the sample) are then utilized in weighting the sample results in order to estimate the characteristics of the population.

The current data collecting efforts at Project TALENT are primarily concerned with the follow-up studies of the students tested in 1960. Each year, about 100,000 people (one of the four grade samples) are sent questionnaires to determine their current jobs and career plans and the amount of education and training they have received since they were tested in high school. The schedule of follow-up surveys is summarized in Table 1.4. Our problem faced by this and similar surveys is how to handle the nonrespondents. People who take the trouble to answer mailed questionnaires may be different from those who choose not to do so or who cannot be located by the post office. To offset this problem of nonrespondent bias the Project TALENT staff selects a sample (one in twenty) of the nonrespondents and locates them through follow-up coordinators living in different regions of the country. These field workers locate and interview these "reluctant respondents" or "hard-to-finds." Their data are weighted by 20 (since there

Table 1.4 Schedule of Project TALENT Follow-Up Surveys

Grade in 1960	1 Year Follow-Up	5 Year Follow-Up	10 Year Follow-Up	20 Year Follow-Up
12th	1961	1965	1970	1980
11th	1962	1966	1971	1981
10th	1963	1967	1972	1982
9th	1964	1968	1973	1983

are 19 others like them who did not respond and were not included in the special field study) and combined with the data from the respondents for analysis.

We tell you all this to emphasize the importance in surveys of having the samples as representative of the population as possible. The TALENT researchers wish to talk about the population of high school students in the United States in 1960 and what they did following high school. If they did not determine and adjust for nonrespondent bias they would be presenting a distorted picture of many important phenomena, even though the initial sample selected in 1960 was unbiased.

1.4 The punched card

We hope that this description of Project TALENT has given you some insight into the nature and purposes of this major educational survey. Another reason for this discussion of TALENT is to provide background for the presentation of two tables of actual survey data. Appendix B contains scores on twenty variables for 234 boys and 271 girls selected from the twelfth-grade TALENT sample. These represent data for every 200th student (0.5 percent samples) of the total twelfth-grade female and male files in the TALENT data bank. The twenty variables have been selected to demonstrate several types of psychometric and sociometric scales (a matter discussed in Chapter 2). When you look at the table in Appendix B you are looking at real, live educational research data, in the raw. They don't tell you much, do they? All they dramatize is the absolute need for statistical reduction and interpretation. The chapters ahead are going to report some statistical analyses on these data and challenge you to perform and interpret some analyses of your own on them. Fortunately, you will have access to computer programs which will do the arithmetic chores for you.

Before such data can be submitted to a computer, they must be punched into machine-readable cards. Figure 1.1 is a reproduction of such a card. Each card contains eighty columns, thereby allowing up to eighty characters (numbers or letters) per card. The card columns are numbered from 1 through 80 under the row of zeros and under the row of 9's at the bottom. Characters are punch-coded in each column. If column 1 is to represent the number 5, a hole is punched in column 1 where the 5 is printed. The card-reading machine senses the hole at the "5 row" for column 1 and "reads" a 5 into the computer. Five "what" depends upon what information was originally punched into column 1. That is, the card designer decides what kind of information is to go into each column. He may decide that the first three columns will be used to store an identification number of the student. A set of columns with such a specific designation is called a field. Thus if a 5 is punched in column 1, a 3 in column 2, and a 6 in column 3, then this

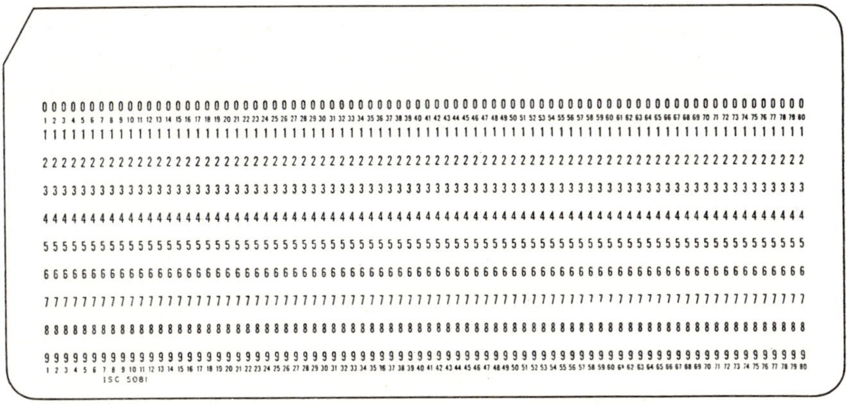

THE PUNCHED CARD 15

Fig. 1.1 The eighty-column punch card.

particular card would contain data for student number 536. These three columns represent the identification field. If columns 11 and 12 are designated to contain the student's age in years, and a 1 is punched in column 11 and a 7 is punched in column 12, then this would indicate that student 536 is 17 years old.

Holes in the card are indicated in Figure 1.2 by rectangular black marks. The character coded by holes in a column is printed by the punch machine at the top of the column for the human reader. Notice the relationship between the hole or holes used to code the character for machine reading and the

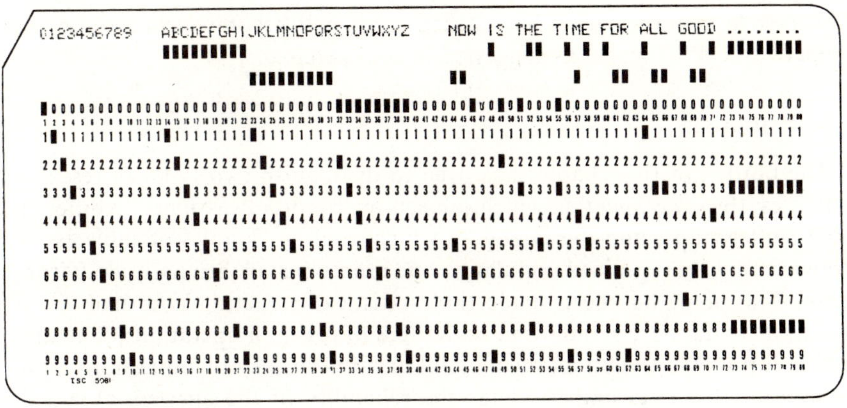

Fig. 1.2 Punched card showing relationship between characters and punches.

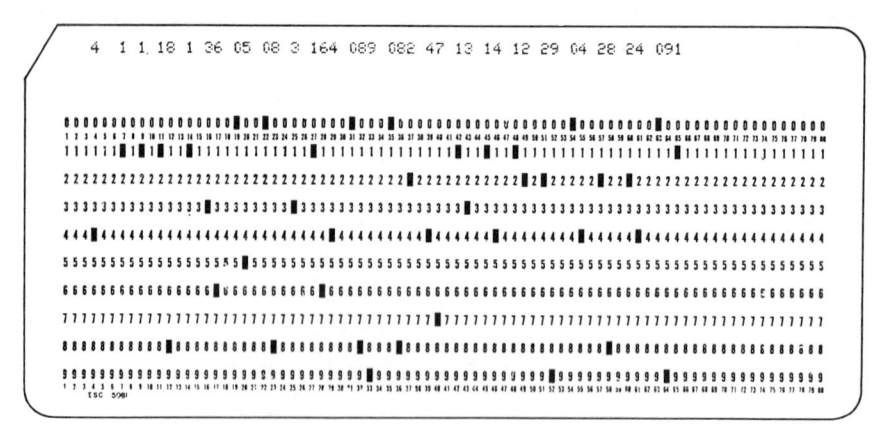

Fig. 1.3 Punched card for student 004.

printed character at the top of each column. The combination of two punches used to code the alphabetic characters in each column can also be seen.

Figure 1.3 is a reproduction of the card punched for the fourth TALENT male in Appendix B. Compare the card with the scores listed for him in Appendix B. The holes punched in cards correspond to the different scores and are the basis for machine reading of the cards. The card layout at the beginning of Appendix B tells what information is punched in each of the columns. The same variable goes in the same column or columns for each student. Thus, when the machine "reads" columns 16 and 17 it will be reading the career plan for the student corresponding to that card. The three-digit student identification number is indicated in columns 2, 3, and 4. This record of information in punched cards will become clearer after you prepare a deck of cards, as suggested in the exercises.

1.5 Summary

The burden of this chapter has been to distinguish two types of research activities, the *experiment*, in which subjects are randomly assigned to different treatment groups, and the *survey*, in which random samples of subjects are drawn from some specified population and are studied in their natural environment. The point has been made that statistical reduction and analysis of quantitative data are standard in both types of research operations. Statistics makes it possible to extract meaningful answers to scientific questions from swarms of raw data and to gauge the reliability of the answers obtained. Finally, the punched card was introduced along with the Project TALENT data of Appendix B.

EXERCISES AND QUESTIONS

1. Your instructor may arrange a demonstration of a card punch machine. Then each member of the class can punch twenty or twenty-five score records for the TALENT data of Appendix B. When the contributions of the entire class are brought together, the class will possess a deck of punched cards suitable for computer processing. You will have acquired a little understanding of how data are prepared for electronic digital computing. In actual research operations, most card punching is done by clerks, or automatically from test answer sheets or cards, but from this experience of preparing the cards you will be more aware of the problems involved in getting *accurate* data decks.
2. What is the major difference between an experiment and a survey?
3. Look through a volume of a research journal (for example, *Journal of Educational Psychology*) and classify the empirical studies reported there as either experiments or surveys. You might wish to come back to this volume at the end of the course and see if it seems less awesome!
4. Look through another research journal (for example, *American Educational Research Journal*) and see if you can find any articles that did *not* present results of statistical analysis methods. How did you decide whether they did or did not?
5. Students should be aware of the fact that different textbook authors emphasize different aspects of their field. We recommend that you parallel your study of this text with frequent excursions into other books. At this point you might examine F. N. Kerlinger (1964). Kerlinger provides an excellent exploration of basic issues in research. He also gives many examples from actual published research.
6. In what important ways does the HOSC experiment differ from the static model described in the third paragraph of Section 1.1?

CHAPTER TWO

Describing Univariate Distributions

2.1 Sources and types of data

The business of behavioral scientists is the establishment of increasingly general theories of human and animal behavior. The theories they produce are abstracted and simplified explanations of behavior which render behavior understandable and predictable. Although theories are merely systems of symbols produced by man's intellect, they are tied to the nature they represent in the following two ways. First, theories are based upon observations of behavior. By a process of induction, the scientist builds theories to explain what he has observed. Second, the new theories then suggest other behavioral phenomena, previously unobserved. If new observations are consistent with those hypothesized from the theories, once again the theories are tied to nature. Observations of behavior and its environment provide the *data* that the scientist organizes and interprets through the use of statistics. In modern laboratory experiments and field surveys the scientist employs a rather extensive assortment of instruments which extend his own powers of observation. In fact, he often collects his data quite automatically. Most of these instruments are designed to read out in digits, and even when the scientist uses himself as the observing instrument, it is customary for him to record his observations numerically, or later to translate verbal or other types of records into numerical form.

Statistical methodology, since it is a branch of applied mathematics, can only be helpful to the scientist who is willing to record his data in numerical form. It is a tribute to the value of statistics that scientists generally quantify their data, even at considerable effort and expense. This is not the place to explore in detail the problems of quantification of data in the behavioral sciences. Solutions to the problems of quantification comprise a well-developed subject called *measurement*, and anyone who takes the trouble to study an introduction to statistics will certainly be willing to undertake an introductory study of measurement separately. In the practice of behavioral science, measurement and statistics are intimately interrelated, but it is well to separate them for initial study. Interestingly, measurement is prior to statistics in practice, in that in designing and executing research the measurement problems are encountered and solved before statistical problems arise. On the other hand, it is easier to *study* statistics before and in preparation for studying measurement, since statistical concepts are involved in the study of measurement. At this point a few comments on measurement are necessary because the general nature of the measurements used determines the kind of statistical method appropriate for analyzing the data resulting from those measurements.

Data may be classified into four types of scales, *nominal, ordinal, interval,* and *ratio*. There are special statistical methods for each type. *Nominal* data are produced by assigning observations to unordered categories. Examples are the two categories for sex: "male" and "female"; or five categories of religion: "Catholic, Jewish, Protestant, Other, or No Religious Affiliation." Measurement in such cases consists of determining in which category an observation belongs. Quantification is achieved by counting the number of observations assigned to each category or group. This results in the group frequencies. It is usually informative to transform these frequencies to proportions or percentages. Nominal scales may be used to code public information about human subjects, such as sex, race, political affiliation, occupation, nationality, and marital status, and questionnaire information on personal preferences. Although numbers may be used to label or name the scale categories, no order is implied by the numbers assigned each category.

Ordinal scales are those on which observations are ranked from "most" to "least," such as socioeconomic level or level of education attained. Again, observations are placed into categories, but these categories are ordered in some way, and are numbered according to their order. Now each observation can be given a *score* corresponding to the order-number of its category. Comparison of scores for two observations allows the statement that one observation has more of the quality underlying the ordinal scale, or is higher on the scale, than the other, unless both scores are the same, indicating that the two observations are tied with respect to the quality. For example, numbers 1 through 5 might be assigned to student responses regarding their educational plans:

1. Leave before completing high school.
2. Finish high school.
3. Some further education or training beyond high school.
4. Finish four years of college.
5. Attend graduate school.

The higher the score, the more education is planned. Thus, the score on an ordinal scale conveys more information about an observation's position relative to other observations than does a "score" on a nominal scale, which merely places the observation *in* one category and *out* of others. It is also useful to count the frequencies of assignments to positions on an ordinal scale. A special case is a complete ranking with no ties of a set of observations, so that the frequency at each position on the scale is one. In this text this special case of ordinal data will be called *ranked* data.

For ordinal scales there is no claim that the distance between one pair of adjacent categories is the same as the distance between any other pair of adjacent categories. Although the categories are usually numbered serially, there is no reason to believe that the difference between two serial numbers, or ranks, is constant in terms of the quality with respect to which the categories have been ordered. For example, a coach may rank his six pitchers, but we cannot assume that in his view the quality difference between the pitchers he ranks 1 and 2 is the same as the quality difference between 2 and 3, or between 5 and 6. Starting with pitcher 6 and moving serially up to pitcher 1, we may assume that the coach believes each pitcher is a better pitcher than those ranked below him, but we cannot say how much better.

When there is a constant unit of measurement between equal intervals on a scale, for all locations along that scale, it is an *interval* scale. This represents the highest type of scale usually achieved in psychological measurement.

Most of the interval-type scales in use are so classified because the investigator is willing to *assume* a constant unit of increment in his scale. In the *Test on Understanding Science* (TOUS), referred to in Chapter 1, each item in the test poses a problem of interpretation of a science situation. A correct solution of any item increments the student's TOUS score by 1 unit, or 1 point. The scale score for a subject is the number of test items he has answered correctly, and it is assumed that the possible scores on the test form an interval scale. For example, the difference of 10 points between two TOUS scores is treated the same regardless of whether the two scores were 30 and 20 or 90 and 80. A 10 point difference is seen as a 10 point difference regardless of where it occurs. A little reflection reveals that there are a large number of combinations of items passed by one subject and failed by another which could underlie a 10 point difference between their total scores. All items would have to have equivalent value as units of understanding of science to completely justify the comparison of the different possible total scores.

Another example is an eight-scale battery called *Readiness for Vocational Planning* (RVP) (Gribbons and Lohnes, 1964), in which eight variables are scaled from different subsets of items in a structured guidance interview. The counselor scoring on the scales is required to make subjective evaluations of the things students have said in the interviews. The questions are open-ended, and result in a great variety of statements that must be quantified by the counselor. Eventually each student receives a set of eight RVP scale scores, and conventional interpretations of differences among students' scores on any one of the scales is encouraged for research purposes. Such interpretations assume that the scales are interval in nature.

The most desirable type of scale, the *ratio* scale, which possesses an absolute zero as well as a constant unit, is seldom obtained for measures of psychological traits. Examples of two ratio scales which are used in behavioral research are height and weight. The major difference between interval and ratio scales is that the "zero" in an interval scale is quite arbitrary (for example, the centigrade thermometer).

The major purpose of this discussion of scale differences is to make sure that you are aware of the fact that numbers resulting from measurement operations may differ, depending upon the nature of those operations. There are no hard and fast rules which dictate whether a particular set of observations is based upon one type of scale or another. One critical difference is what the investigator is willing to assume about the scale he is using. Of course, the more he can justify the assumptions he does make, the more confidence others will have in his results. The reason the investigator wants to make certain assumptions about his scales is that he may then use statistical methods which are only appropriate if he has the assumed type of scale. This concept will become clearer as you gain more familiarity with the variety of statistical techniques that are available to the researcher.

2.2 The frequency distribution

When the behavioral scientist applies a measurement instrument or procedure to a number (N) of subjects, the result is a set of numbers, usually termed "scores." Often his ultimate purpose is to relate this set of scores to some other set of scores in order to test some hypothesis about the relationship between the two variables represented by the two sets of scores. Later the text will introduce statistical methods for describing relationships between variables and testing hypotheses about such relationships. However, before inquiring about the relation of a set of scores to some other set, the scientist is well-advised to pause and inquire into the characteristics of the obtained set of scores for one variable. The remainder of this chapter is concerned with methods of describing the characteristics of a set of scores, and of modifying those characteristics in desirable ways.

An assured characteristic of every set of scores is variability. In fact, the convention of calling the trait which is measured by an instrument a *variable* dramatizes the expectation that a variety of score values will be assigned to the subjects by the instrument, corresponding to a variety of values or strengths of the trait possessed by the different subjects. Of course, "a variety" in some cases means a minimum of two values, when the variable allows only two states, as when it is the sex of subjects. In other cases, such

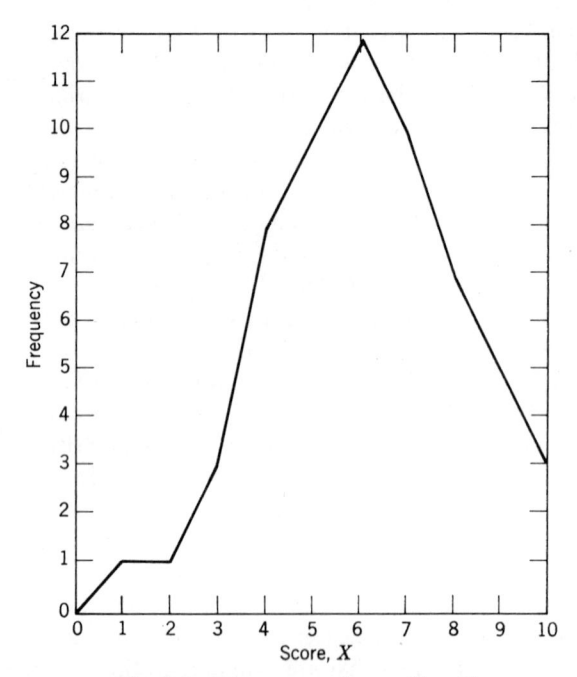

Fig. 2.1 Frequency polygon, $N = 60$.

as the age or income of subjects, there may be a large number of possible values. Whether the scores are assigned in terms of a few states or "categories," or as a large number of "points" on a scale, it is informative to count the *frequency* of assignment of each category or point. These frequencies may be reported either in tabular or graphical fashion to make the variability of the data visible. Table 2.1 and Figure 2.1 represent these alternatives for

Table 2.1 Frequency Distribution

Score (X)	Frequency (f)
1	1
2	1
3	3
4	8
5	10
6	12
7	10
8	7
9	5
10	3
	$N = 60$

THE FREQUENCY DISTRIBUTION

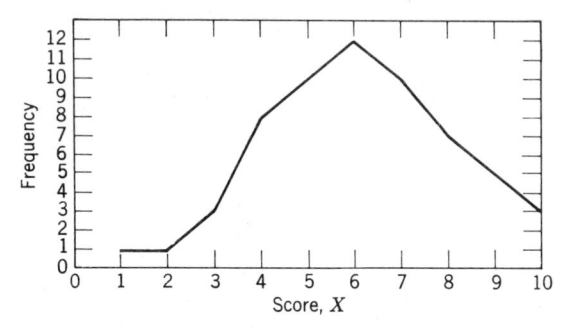

Fig. 2.2 Frequency polygon, $N = 60$.

reporting the same set of frequencies. Both are reports of a frequency *distribution*, but the graph is further designated as a "frequency polygon."

Graphs are generally more meaningful than tables. However, the possibilities of manipulating the length of vertical units for the frequency and the horizontal units for the scale values to produce different impressions of the frequency distribution (compare Figures 2.1 and 2.2) make it necessary for the reader to be wary of the unethical graph maker. For an entertaining discussion of the many different ways in which graphical methods can be used to distort the facts, see *How to Lie with Statistics*, by Darrell Huff (1954).

Figure 2.3 demonstrates another graphical report of the distribution of the scores, called a "histogram." It, too, can be given a variety of shapes by changing the vertical and horizontal scales.

When the number of score points, or discrete values of the measurement, is large, it is useful to group adjacent points into categories and to count and

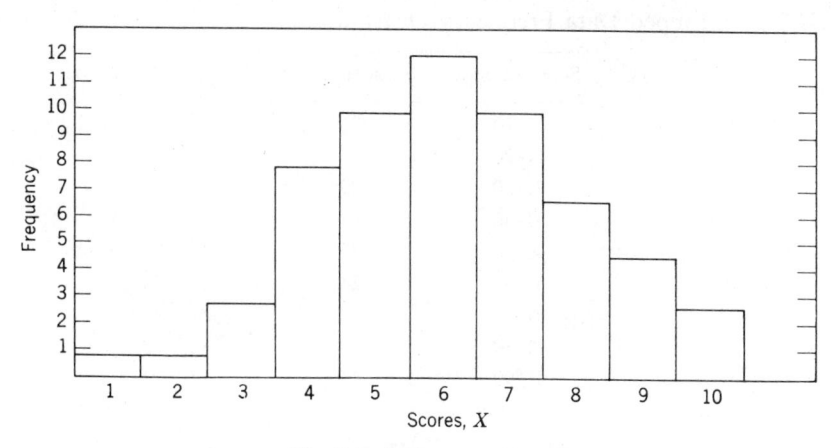

Fig. 2.3 Histogram, $N = 60$.

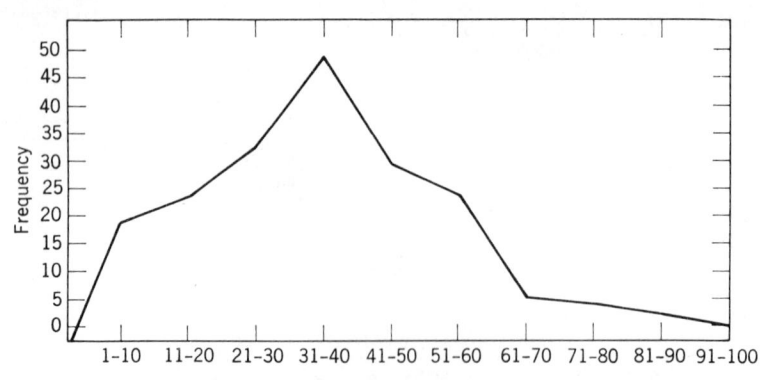

Fig. 2.4 Frequency polygon, $N = 200$.

report the frequency of subjects in each category. Table 2.2 and Figure 2.4 illustrate a grouped data frequency distribution. Of course, the choice of an interval defining the number of score points grouped in a category and the total number of categories employed is quite arbitrary, so once again there is the possibility of manipulating the appearance of the frequency distribution. Note that every table or graph of a frequency distribution should report the total number of subjects (N) whose scores are distributed.

Sometimes it is useful to table the proportion (p) of N assigned to each score category. If we use the symbol N_j for the number of cases in category j, and p_j for the proportion of cases in that category, then

$$p_j = \frac{N_j}{N} \tag{2.2.1}$$

Table 2.2 Grouped Data Frequency Distribution

Score Interval	Frequency (f)
1–10	22
11–20	25
21–30	33
31–40	47
41–50	31
51–60	24
61–70	8
71–80	5
81–90	3
91–100	2
	$N = 200$

THE FREQUENCY DISTRIBUTION

Table 2.3 Proportion Receiving Each Score, $N = 60$

Score	f	Proportion
1	1	.017
2	1	.017
3	3	.050
4	8	.133
5	10	.167
6	12	.200
7	10	.167
8	7	.117
9	5	.083
10	3	.050
	60	

Another attractive approach is to table the percentage (%) of N assigned to each category, where percentage equals 100 times the proportion:

$$\%_{oj} = 100 p_j \tag{2.2.2}$$

It is especially important that tables of proportions and percentages report the actual total N. Table 2.3 reports the frequency distribution of Table 2.1 in terms of proportions. For example, $p_4 = 8/60$ which is tabled as .133. Table 2.4 reports the frequency distribution of Table 2.2 as proportions and as percentages. A very interesting thing about the tables of proportions is that at times they may be interpreted as tables of probabilities. In this context the probability of an event is the number of times the event occurred in a sufficiently large number of trials divided by the number of times it could have occurred. In an observed set of events, it is easy to interpret the proportion receiving the jth score as the probability of a single subject's

Table 2.4 Percentage in Each Score Interval, $N = 200$

Score Interval	f	Proportion	Percentage
1–10	22	.110	11.0
11–20	25	.125	12.5
21–30	33	.165	16.5
31–40	47	.235	23.5
41–50	31	.155	15.5
51–60	24	.120	12.0
61–70	8	.040	4.0
71–80	5	.025	2.5
81–90	3	.015	1.5
91–100	2	.010	1.0
	200	1.000	100.0

receiving that score. If this seems unreasonable, suppose that Table 2.3 reports the proportions of N subjects who received each of the ten possible scores on a test, and suppose that I am one of the subjects. I refuse to tell you which one; nevertheless, I demand to know what the chances are that I received a maximum grade on the test. If I press you enough, will you tell me that the probability of receiving a grade of 10 is .05? If not, why not? This simple notion that under some circumstances the proportions based upon an observed frequency distribution of a variable X can be taken as the probability distribution for X is one of the basic ideas of statistics. What these circumstances are will be discussed later.

In summary, the observed frequency distribution permits you to visualize a set of scores in either tabular or graphical form. The entries in a tabled frequency distribution may be frequencies, proportions, percentages, or probabilities, and the data may be grouped when desirable.

A variation on these methods of displaying a score distribution is the *cumulative frequency distribution*. In this presentation each score has associated with it the frequency of subjects having that score *and* all scores lower than that score. When data are grouped in intervals, the cumulative frequency for a given interval is the sum of the frequencies for all intervals up to and including that one. It is also common practice to table *cumulative proportions*, rather than frequencies. The cumulative proportion for a score, X, may be interpreted as indicating the probability of a randomly selected subject's score falling below or at X. The difference $1 - p$ may be interpreted as the probability of a random observation falling above X. Cumulative frequencies or cumulative proportions are plotted in a graph called an "ogive." Table 2.5 reports the cumulative frequencies and cumulative proportions for the grouped data of Tables 2.2 and 2.4, and Figure 2.5 is the graph of the ogive for these data.

Table 2.5 Cumulative Distribution for Grouped Data of Table 2.2, $N = 200$

Score Interval	Frequency	Cumulative Frequency	Proportion	Cumulative Proportion
1–10	22	22	.110	.110
11–20	25	47	.125	.235
21–30	33	80	.165	.400
31–40	47	127	.235	.635
41–50	31	158	.155	.790
51–60	24	182	.120	.910
61–70	8	190	.040	.950
71–80	5	195	.025	.975
81–90	3	198	.015	.990
91–100	2	200	.010	1.000

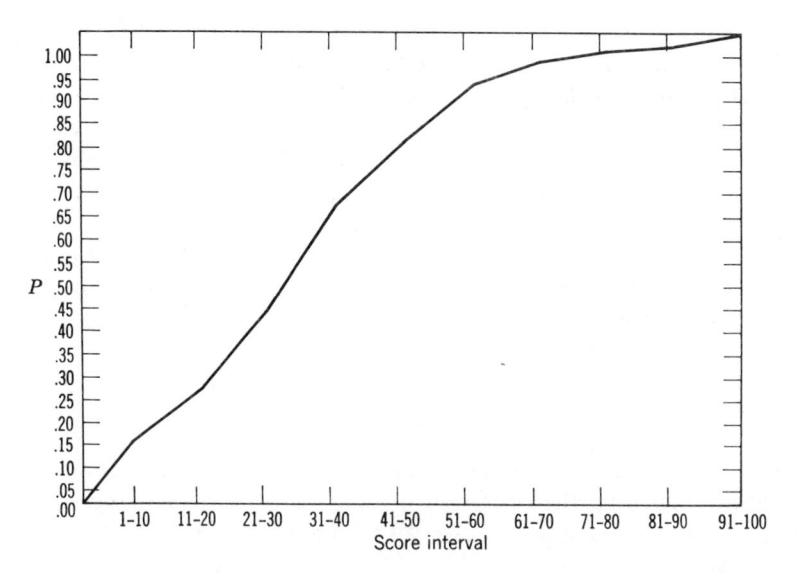

Fig. 2.5 Cumulative ogive, $N = 200$.

2.3 Central tendency

All the frequency distributions pictured in the previous section had only one peak. The peak score having the largest frequency is called the *mode* of the frequency distribution. Distributions with one peak or mode are called *unimodal*. Occasionally the mode is the best report of the typical performance, or central tendency, in a group of subjects. Usually it is more informative to know the *median* score, which is the point on the score scale dividing the top half of the distribution from the bottom half, that is, the score value above which 50 percent of the subjects scored. The median is found by ranking the obtained scores from lowest to highest and counting down $(N + 1)/2$ scores from the highest. For the distribution reported in Table 2.1 and Figure 2.1, the mode is 6 and the median is 6. The median is perhaps the most easily interpreted statistic for reporting the central tendency in a distribution, and it is sometimes preferred, as when the distribution contains a few extreme cases. Usually the statistician urges the employment of the arithmetic average, or mean, because it can be integrated with the general theory of statistics more readily and completely than can the median. For essentially technical reasons this text will stress the mean as the estimate of typical performance. The mean is only appropriate, however, if you are willing to assume that you have an interval or ratio scale.

The *mean*[1] is the sum of scores divided by the number of scores. If X_1 is the score of the first subject, X_2 the score of the second subject, and so on to X_N, the score of the last subject, the mean of X, designated M_x, is defined as

$$M_x = \frac{X_1 + X_2 + \cdots + X_N}{N} \tag{2.3.1}$$

If X_i stands for any one of the terms inside the numerator, that quantity which is in the numerator represents the sum of the X_i, where i takes the values of the integers from 1 to N. As shorthand for the numerator term statisticians employ the notation

$$\sum_{i=1}^{i=N} X_i$$

The symbol $\sum$, which is a capital sigma, is ordinarily used in this way as an "operator" signifying the operation of adding, with the *subscript* (in this case i) on the *variable* (in this case X) taking a set of values in the *range* (in this case 1 to N).

For the data of Table 2.1, the sum of scores is 367 and the mean is 6.1. If we had the complete roster of 200 scores underlying Table 2.2 it would be possible to sum them by hand to get their mean, but perhaps the reader will agree with the authors' preference for disposing of such arithmetic chores by mechanical or electrical machine methods.

2.4 Statistics and parameters

Since the mean is a very basic and heavily used statistic, it is an appropriate referent for a definition of the term "statistic." Under what conditions is a number a statistic? In everyday language there is a very general meaning in which almost any number is a statistic. Statisticians, however, use the term to refer to a special class of numbers. To understand their usage of the term, you need to know the distinction between a "population" and a "sample." In statistics, a *population* is a class of events (frequently but not always people) that the scientist seeks knowledge of. A population may be theoretically infinite in size (for example, all mothers of twins), or it may be finite but large in size (for example, all registered Democrats in 1966). Presumably there are certain descriptive numbers that would characterize the distribution of scores on a variable for all the members of the population, if only these descriptive numbers were known. For example, the scores on variable X for a population have, in theory, a mean. This population mean is called a

---

[1] The authors have chosen to use M as a symbol for the mean because of its value as an acronym for "mean" and its parallelism to μ, the standard symbol for the population mean. Some texts use $\bar{X}$ (X-bar) as a symbol for the "mean of X."

parameter for the population. We will use the Greek μ (mu) for population means. Thus, μ_x represents a population mean for variable X. This is a theoretical quantity, since it is not known to the scientist. What the scientist does is collect measurements on what he hopes is a representative sample of the population and then compute the mean for the sample, which we call M_x, as an estimate of the population mean, μ_x. To statisticians, a *sample* is a subset of a population drawn in such a way that every member of the population has a known, nonzero probability of appearing. Techniques and theory of sampling are the concern of a vast literature in statistics, but it suffices to say here that every statistical sample incorporates an element of random selection. Randomness in selection occurs when it is impossible to predict exactly which members of the population will appear in the sample, so that every member has some chance to be drawn. Project TALENT was studying a 5 percent sample of American high school students in 1960. Its subjects were selected by a sampling scheme in which randomization played a part, so that every American high school student in 1960 had a known, nonzero chance of appearing in the sample.

A statistic is an estimate of a population parameter computed on a sample. For example, M_x is a statistic that estimates, from the scores on X of a sample of N subjects, the population parameter μ_x. Statisticians say that M_x describes the central tendency in a sample and estimates the central tendency in the population. The sample mean is descriptive of the sample, but far more importantly, it provides an inference regarding the population. It is customary in statistics to denote parameters with Greek letters and statistical estimators of parameters with corresponding Roman letters.

A typical use of the mean as a descriptive statistic occurred in research on the validity of a statewide testing program. In one aspect of this project, attention was focused on the test battery performances of high school sophomores in two different school systems. Since the validity question under study was the ability of the tests to predict curriculum membership, students in each high school were separated into "college" and "noncollege" curriculum groups. Table 2.6 reports the means on the six tests for each curriculum group in each school. Evidently the college curriculum group scored, on the average, above the noncollege group on every test in both schools, which suggests the existence of the sought-for validity. Furthermore, on every test the college group of School 1 scored above the college group of School 2, and the noncollege group of School 1 was superior to that of School 2, but to a lesser degree. Thus, the table of means seems to support the claim that the test battery can discriminate among curriculum groups within schools. Actually, you will soon realize that other considerations are involved in a statistical test of this claim.

Table 2.7 reports the means on the same tests achieved by a sample of the

Table 2.6 Sophomore Test Battery Means for Two Curriculum Groups in Two Schools[a]

	School 1		School 2	
Variable	College ($N = 239$)	Noncollege ($N = 204$)	College ($N = 206$)	Noncollege ($N = 156$)
SCAT[b] Verbal	293	276	283	274
SCAT Quantitative	305	293	295	289
COOP[c] Reading Vocabulary	159	149	154	149
COOP Reading Level	156	146	152	145
COOP Reading Speed	159	149	152	146
COOP Reading Expression	158	146	153	145

[a] From P. R. Lohnes and P. H. McIntire (1963).
[b] SCAT is the *School and College Ability Tests*, Princeton, N.J.: Educational Testing Service, 1955.
[c] COOP is the *Cooperative English Tests*, Princeton, N.J.: Educational Testing Service, 1960.

sophomores in all the public high schools in the state, again for two curriculum groups. Such statewide sample means may be called "norms," and illustrate another popular use of means. If the individual school means

Table 2.7 Sophomore Test Battery Means for a Stratified Random Sample from All Public High Schools in New Hampshire[a]

Variable	College Preparation ($N = 404$)	Noncollege ($N = 424$)
SCAT Verbal[b]	289	275
SCAT Quantitative	302	288
COOP[c] Reading Vocabulary	156	148
COOP Reading Level	155	146
COOP Reading Speed	156	146
COOP Reading Expression	155	146

[a] From P. R. Lohnes and P. H. McIntire (1963).
[b] SCAT is the *School and College Ability Tests*, Princeton, N.J.: Educational Testing Service, 1955.
[c] COOP is the *Cooperative English Tests*, Princeton, N.J.: Educational Testing Service, 1960.

of Table 2.6 are compared with the statewide norms of Table 2.7, it appears that School 1 is consistently a "superior" school, while School 2 is a "below average" school. The characterizations apply more to the college preparatory divisions of the schools in this instance.

The mean of a distribution of scores characterizes the central tendency, or center of gravity, of the group's performances, It does not convey any sense of the extent of variability in those performances, however. For example, Table 2.8 reports the test performances of ten subjects in each of two groups.

Table 2.8 Test Score Rosters for Two Groups of Ten Subjects Each

Subject Number	Group 1 Scores	Group 2 Scores
1	3	1
2	4	2
3	4	3
4	5	3
5	5	4
6	5	5
7	5	6
8	5	6
9	7	9
10	7	11
	$M_1 = 5.0$	$M_2 = 5.0$

The group means are the same, but clearly Group 2 is much more variable in performance than is Group 1. We sometimes say that Group 2 is more "heterogeneous" than Group 1 (or alternately, that Group 1 is more "homogeneous" than Group 2). This example demonstrates the need for a statistic to describe the extent of variability in a set of scores, which is the subject of the next chapter.

2.5 Summary

Four types of scales used in measurement by behavioral scientists are as follows.

1. *Nominal* scale (unordered categorical data).
2. *Ordinal* scale (ordered or ranked categories).
3. *Interval* scale (ordered data with a constant unit of distance between equal intervals on the scale).
4. *Ratio* scale (interval scale with a meaningful zero).

Scores for a sample of subjects on any of the types of scales can be displayed in a *frequency distribution*. The display may take the form of a *polygon*, *histogram*, or *table*. The graph of a cumulative distribution is an *ogive*.

For any type of scale, the *mode* locates the most frequently assigned score. For ordinal, interval, or ratio scales, the *median* locates the middle or central score in the distribution. For interval and ratio scales, the *mean* locates the center of gravity, and is the measure of central tendency most frequently employed by behavioral scientists. The mean is the sum of the scores divided by the number of them. This statistic is defined algebraically as

$$M_x = \frac{\sum_{i=1}^{i=N} X_i}{N}$$

The sample statistic M_x is an estimator of the unknown population parameter μ_x.

Since two samples of subjects may have the same N and the same mean, yet differ markedly in the extent of scatter around the mean, there is need for a measure of the amount of scatter as a second characteristic describing a frequency distribution. The next chapter considers this problem.

EXERCISES AND QUESTIONS

1. Turn to Appendix B and classify the twenty TALENT variables listed there into each of the four categories of scales. Reconciling differences with classmates should produce good class discussion!

2. What is the mode for variable 3 (age) of the first 48 boys listed in Appendix B?

3. A student who wished to describe the typical career plan of twelfth-grade males found and reported the mean of variable 5 in Appendix B. (a) What is wrong with doing this? (b) How might the student accomplish his objective?

4. Determine the frequency distribution for the sociability score (variable 17 in Appendix B) for the first 48 girls in Table B.3. Construct a frequency polygon and a histogram for this distribution. Isn't it easier to think about a particular student's score once you have "the big picture" regarding the distribution?

5. It might be said that the first column of means in Table 2.7 describes the typical college preparatory student in New Hampshire public high schools. What is the sense of this assertion? What might be the nonsense?

CHAPTER THREE

Variability in Measurements

3.1 Variability

One of the most popular and successful branches of psychology is called "the study of individual differences." Of course, it doesn't require a psychologist to note that people differ. One of the writers of the Bible reported on the individual differences he found in a single pair of brothers, Cain and Abel. The extensive differences among individuals is one of the most basic conditions of human experience.

Early in the history of the scientific study of human behavior, psychologists began devising tests for observing and quantifying facts of individual differences. One strategy for studying differences is the comparison of mean performances of different groups of people (for example, between sexes) or the comparison of mean performances of a special group of people with norms for the population as a whole. This method of *mean differences* has already been illustrated using Tables 2.6 and 2.7. In his famous *Genetic Studies of Genius*, Louis Terman's method was to select a special group of youths whose general intelligence test scores were high enough to place them in the upper 2 percent of their age group with respect to performance on the test, to study the lives of this sample over the years as they developed, and to compare their mean achievements in many areas with the population averages. It is not surprising that Terman was able to show superior educational attainments for his "geniuses," but their outstanding achievements in most other areas did surprise some people, and the research has helped to destroy the folklore about academically gifted individuals being unbalanced individuals.

Another strategy of research on individual differences focuses on the distribution of differences on one or several traits *within* a single group, rather than on differences *among* groups. The mean score for a group gives information about the "center of gravity" around which individual scores are distributed, but it does not tell a thing about the extent of observed differences. Inspection of Tables 2.6 and 2.7, which report only means, indicates that such reports leave the reader totally uninformed about the scatter or variability of performances in each of the groups. However, a little thought about the possible degrees of variation within the groups leads to the conclusion that *the importance to be attached to the differences among means cannot be decided in the absence of information about group variabilities.*

That conclusion can be illustrated by use of Table 2.6, which shows a mean SCAT Verbal score of 293 for college preparatory students in School 1 and a corresponding mean of 283 for School 2. Suppose both groups are very *homogeneous* (show little variation) in performances on this test, so that *all* SCAT verbal scores in the first group lie in the range 284 to 302, and *all* scores in the second group lie in the range 274 to 292. In such a situation, no college preparatory student from School 1 scored as low as the mean for prep students of School 2, and no prep student from School 2 scored as high as the mean for prep students of School 1. This situation is displayed as Case I in Figure 3.1. There is some overlap of the two frequency polygons, but the separation of the means is very significant. On the other hand, suppose both groups are very *heterogeneous* (show large variation) in performances on the test, so that many Group 1 prep students scored below the Group 2 mean of 283 and many Group 2 prep students scored above the Group 1 mean of 293. Case II in Figure 3.1 illustrates this possibility. Now the difference between the group means appears to be much less significant.

Another illustration of the previous generalization about mean differences begins with the question of how much information a student's test score conveys about the probable group membership of that student. In Case I it is apparent that if the score is at or above 293 the student must belong in Group 1, and if the score is at or below 283 he must belong in Group 2. It is only in the range 284 to 292 that it is necessary to guess the group membership of the score's owner. Even in this range, scores from 284 to 287 may be said to be much more likely to belong to Group 2, and the scores from 289 to 292 may be said to be much more likely to belong to Group 1. It is only for a score of 288 that a coin must be flipped to produce a guess about group membership.

Turning to Case II, it appears that although scores above 288 are more likely to belong to Group 1 and scores below 288 are more likely to belong to Group 2, there are few scores for which the student's group membership is certain, and guesses about the memberships of owners of most scores (except 288) are much more hazardous than in Case I. Thus the two cases support the generalization that the greater the variability within groups the less important is a given mean difference among groups.

At this point, consideration of Table 2.8 suggests another generalization, namely that *two groups may differ in variability of performances even if there is no important mean difference.* This suggests that any research strategy must

VARIANCE

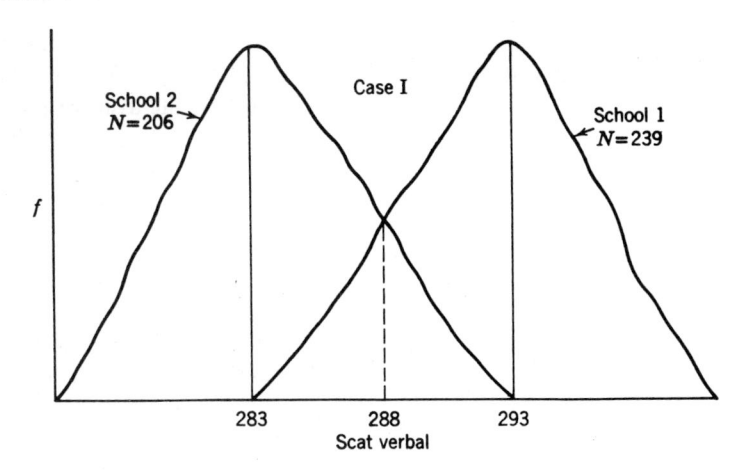

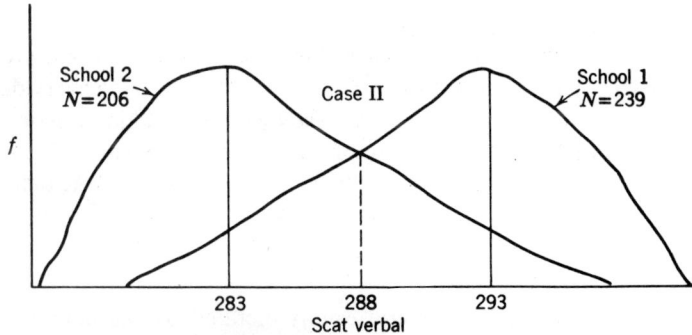

Fig. 3.1 Hypothetical frequency polygons for SCAT verbal scores for college preparatory students from two schools.

pay attention to differences between group variabilities as research findings in their own right, as well as for assistance in interpreting mean differences. It should now be clear that a statistic for describing the magnitude of group variability is quite necessary.

3.2 Variance

You may already have concluded that a useful measure of variability in a group is the extent to which individual scores deviate from the mean of the group. That is, we can find a *deviation score* for each individual and average these deviation scores. For example, let us take one individual, individual i whose observed score is X_i. This person's deviation score (x_i) is:

$$x_i = X_i - M_x \tag{3.2.1}$$

Note that x_i carries a negative sign if the person's score is below the mean. Thus the sign of a deviation score tells immediately whether the observed score was above or below average. It might seem that the average deviation score would be a reasonable statistic to describe the extent of variability in a score distribution. But consider what the following algebra tells us about the sum of any set of score deviations from their common raw score mean.

$$\begin{aligned}
\sum x_i &= \sum (X_i - M_X) & &\text{by definition} \\
&= \sum X_i - \sum M_X & &\text{by summation rule I[1]} \\
&= \sum X_i - N M_X & &\text{by summation rule II, since } M_X \text{ is a constant for a given set of } N \text{ scores} \\
&= \sum X_i - N \left(\frac{\sum X_i}{N} \right) & &\text{substituting for } M_X \\
&= \sum X_i - \sum X_i & &\text{cancelling the } N\text{'s}
\end{aligned}$$

Therefore, $\sum x_i = 0$

In short, the sum of any set of deviation scores must be zero, creating a useless statistic. A more useful statistic is created by squaring every deviation score and taking the mean of these squared deviations. This statistic is called the *variance*, which is a highly suitable name for a measure of variability. Statisticians identify the variance by the symbol s^2. Its algebraic definition is

$$s_X^2 = \frac{\sum x_i^2}{N} \qquad (3.2.2)$$

Notice that the order of arithmetic events dictated by the definition is (1) square each deviation score, (2) sum these squares, and (3) divide this sum by N, the number of squares. Of course, this definitional equation assumes possession of a set of deviation scores. An alternative and algebraically equivalent formula is

$$s_X^2 = \frac{\sum (X_i - M_X)^2}{N} \qquad (3.2.3)$$

---

[1] Summation rule I: distributing a summation sign

$$\sum (x_i + y_i) = \sum x_i + \sum y_i$$

Summation rule II: summing a constant C

$$\sum_{i=1}^{N} C = NC$$

Summation rule III: summing a constant multiplier of a variable

$$\sum C x_i = C \sum x_i$$

VARIANCE

This equation begins with raw scores but assumes prior computation of the mean. Another equivalent formula is

$$s_X^2 = \frac{1}{N}\left[\sum X_i^2 - \frac{(\sum X_i)^2}{N}\right] \quad (3.2.4)$$

This last is an example of a computing formula, because although it is clearly a clumsier expression to remember, it is an easier way to compute s^2. Note that it requires only N and two accumulations or sums, namely, $\sum X_i$ and $\sum X_i^2$, the sum of raw scores (needed to get M_X anyway), and the sum of squared raw scores.

For small sets of data these accumulations can be done with paper and pencil from the frequency distribution. Using the Table 2.1 distribution, the first step is to produce the f times X column of Table 3.1. The sum of this

Table 3.1 Hand Computation of Variance ($N = 60$)

X	f	$f \cdot X$	$(f \cdot X) \cdot X$
1	1	1	1
2	1	2	4
3	3	9	27
4	8	32	128
5	10	50	250
6	12	72	432
7	10	70	490
8	7	56	448
9	5	45	405
10	3	30	300
	$N = 60$	$\sum X = 367$	$\sum X^2 = 2485$

Substituting these three column sums into Equation 3.2.4:

$$s_X^2 = \frac{1}{60}\left[2485 - \frac{(367)^2}{60}\right]$$

$$= \frac{1}{60}\left(2485 - \frac{134{,}689}{60}\right)$$

$$= \frac{1}{60}(2485 - 2244.82)$$

$$s_X^2 = 4.0$$

resulting third column is the total sum of X. Then multiplying the score X times the corresponding fX in each row the resulting fourth column results in the sum of the squares of X. These accumulations are then substituted into Equation 3.2.4 to obtain the desired variance. Notice that $\sum X^2$ is quite different from $(\sum X)^2$, 2485, and 134,689, respectively. You might have used Equations 3.2.2 or 3.2.3 as the basis for computing this variance but you should be able to see now the advantages of this raw score computational procedure. In Chapter 4 you will learn how the computer can take on these computing chores for you. In (3.2.4) the bracketed term is equivalent to the numerator of (3.2.2) and gives the "sum of squared deviation scores." This "sum of squares," as it is generally called, is encountered very frequently in statistics.

The algebraic derivation of the raw score accumulations expression for this sum of squares is not too difficult to follow, so we present it here to illustrate for you how statisticians go from one algebraic expression to another. We try to keep these derivations to a minimum but we feel that some exposure to derivations is essential if you are to appreciate that statistics is a branch of mathematics.

Algebra — *Description*

$\sum x_i^2 = \sum (X_i - M_x)^2$ — Definition of sum of deviation squares

$\sum x_i^2 = \sum (X_i^2 - 2X_i M_x + M_x^2)$ — Squaring the right-hand expression

$\sum x_i^2 = \sum X_i^2 - \sum 2X_i M_x + \sum M_x^2$ — Distributing summation sign (summation rule I)

$\sum x_i^2 = \sum X_i^2 - 2M_x \sum X_i + \sum M_x^2$ — Since $2M_x$ is a constant multiplier (summation rule III)

$\sum x_i^2 = \sum X_i^2 - 2M_x \sum X_i + NM_x^2$ — Since M_x is a constant (summation rule II)

$\sum x_i^2 = \sum X_i^2 - 2\left(\dfrac{\sum X_i}{N}\right)\sum X_i + N\left(\dfrac{\sum X_i}{N}\right)^2$ — Substituting $\sum X_i / N$ for M_x

$\sum x_i^2 = \sum X_i^2 - 2\dfrac{(\sum X_i)^2}{N} + \dfrac{(\sum X_i)^2}{N}$ — Converting the middle term from a product to the corresponding square and cancelling the N's in the last term

$\sum x_i^2 = \sum X_i^2 - \dfrac{(\sum X_i)^2}{N}$ \hfill (3.2.5) — One of the two $(\sum X_i)^2/N$ in the negative middle term cancels out the positive final term

Now compare this derived expression (3.2.5) for $\sum x_i^2$ with the bracketed

STANDARD DEVIATION

portion of Equation 3.2.4 and you should see the equivalence of (3.2.2) and (3.2.4).

It can be shown that for a given set of data the sum of squares of deviations around the mean is smaller than any other possible sum of squares around any other value you might choose. This condition of "least squares" is often used by statisticians in considering possible statistics. Because it is so important, the following derivation is given to illustrate further the condition of least squares. Imagine some arbitrary value, ϕ (phi), which is not the mean. The chosen ϕ will differ from the mean by a constant c such that $M_x \pm c = \phi$. Now a deviation from ϕ can be expressed:

$\tilde{x}_i = X_i - \phi$	Defines $\tilde{x}_i$ as a deviation of X_i from some constant ϕ
$\tilde{x}_i = X_i - (M_x \pm c)$	Substitutes definition of ϕ for ϕ
$\tilde{x}_i = (X_i - M_x) \pm c$	Rearranging parentheses
$\tilde{x}_i^2 = [(X_i - M_x) \pm c]^2$	Squaring both sides of equation
$\sum \tilde{x}_i^2 = \sum [(X_i - M_x) \pm c]^2$	Summing for all values of i from 1 to N
$\sum \tilde{x}_i^2 = \sum (X_i - M_x)^2 \pm 2c \sum (X_i - M_x) + Nc^2$	Expanding (squaring) right side of equation and applying summation rules
$\sum \tilde{x}_i^2 = \sum (X_i - M_x)^2 + Nc^2$	Removing middle term because sum of deviation is zero, that is, $\sum (X_i - M_x) = 0$

This final equation shows that the sum of deviation squares from any point other than the mean is larger by Nc^2 than the sum of deviation squares from the mean. The latter is therefore a least squares accumulation. This concept of selecting a statistic which minimizes a sum of squares occurs repeatedly in statistics.

3.3 Standard deviation

The variance, as a measure of variability, is one of the most important elements in the theory of statistics, as will become apparent later. It is not particularly useful as a *descriptive* statistic, however. For example, it would be useful to be able to describe the scatter of scores in Table 3.1 by using an index that allowed you to determine what proportion of the scores lies within a given score interval from the mean. The problem with the variance is that it expresses the extent of scatter in the score distribution in units which are squares of the original score units. The mean in Table 3.1 is approximately equal to 6. In that particular distribution, about one-half (32 out of 60) of the scores lie within 1 score unit on either side of the mean. The number of score units from the mean that will include, say, 50 percent of the cases will vary from distribution to distribution. Consider the three frequency distributions and polygons of Figure 3.2. All three distributions have a mean of 5, but they have quite different scatter about that mean. In the bell-shaped distribution, 54 percent of the cases fall within 1 score unit on either side of the mean (between 4 and 6), while only 33 percent and 13 percent do for the rectangular and U-shaped distributions, respectively.

In describing scatter in these examples we have used the proportion of cases that is within a fixed score interval from the mean. A more useful description is to find the number of score units that is needed for a fixed proportion of the cases. For example, how many units on either side of the mean are needed to include two-thirds of the scores in the distribution.

Problems with the variance as a descriptive statistic can be solved by defining the *standard deviation*, s, as the square root of the variance. Thus

$$s_X = \sqrt{\frac{\sum x_i^2}{N}} \qquad (3.3.1)$$

The standard deviation is a measure of variability expressed in the *original units of measurement*, and thus compatible with the mean. In fact, since most units of measurement employed in the behavioral sciences are rather arbitrary, a comparison of the magnitudes of both the mean and the standard deviation is often much more meaningful than is either statistic viewed by itself.

The standard deviation can be used to determine the proportion of scores (observations) we can expect to find within a given interval about the mean. We are indebted to a mathematician named Tchebysheff (Russian, 1821–1894) for a theorem which tells what proportion of the scores in a distribution *must* lie within the limits created by adding to the mean and subtracting from the mean any multiple (k) of the standard deviation (s), where k is a constant greater than 1 ($k > 1$). The theorem states that the limits $M_X \pm ks$ enclose at least the $1 - 1/k^2$ part of N. Figure 3.3 illustrates this for values of k of 1.5 and 4.0. Tchebysheff's theorem makes no assumptions about the shape of the distribution of N measurements and applies to any set of numbers

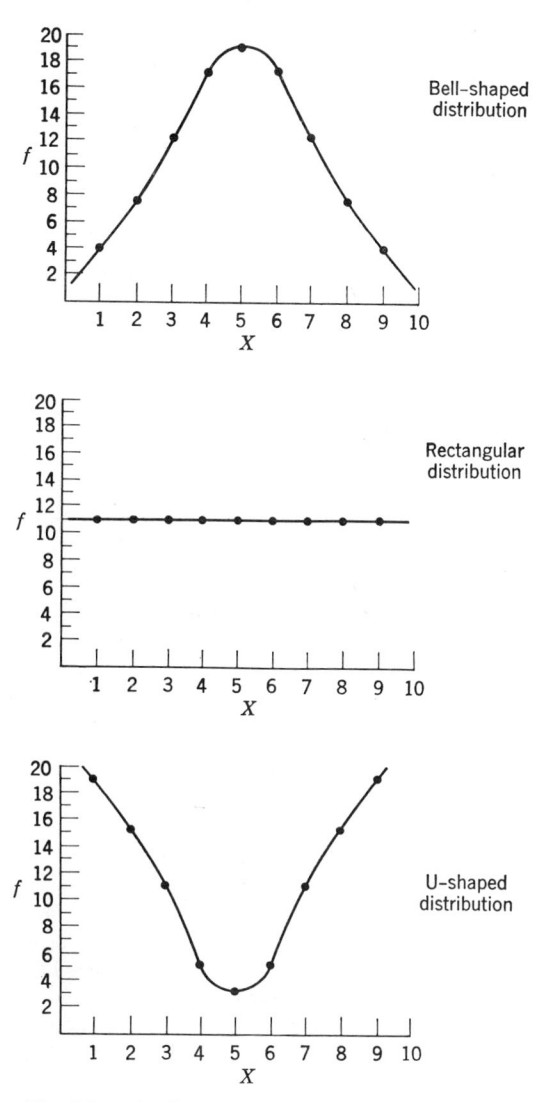

Fig. 3.2 Distributions with three different shapes.

whatever. The theorem assures us that at least 56 percent of all subjects contributing scores to any distribution lie within 1.5 standard deviations from the mean, 75 percent lie within 2 standard deviations, 89 percent within 3, and at least 94 percent within 4 standard deviations from the mean.

An interesting exercise is to test the theorem on the two score rosters reported in Table 2.8. For these two distributions, intervals based on $k = 1.5$

include nearly all the cases, and intervals based on $k = 2.0$ exhaust the cases, so that the theorem is certainly true (in fact is conservative) in these cases. It will be shown later that when the frequency polygon approximates the bell-shaped curve known as the "normal" curve, a much larger proportion of the scores is contained in the interval based on any particular k, so that a stronger rule is possible.

The ratio of a person's deviation score to the standard deviation conveys information about his placement in the distribution. For example, consider a student with a score of 60 in a distribution where the mean is 45 and the

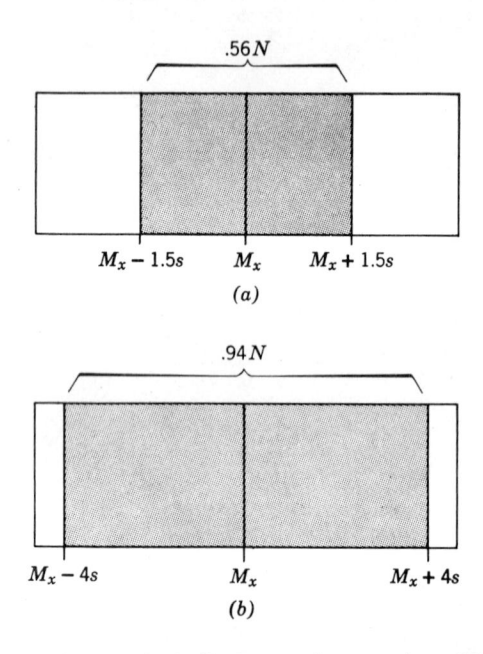

Fig. 3.3 Graphs illustrating Tchebysheff's theorem for two values of k. (*a*) When $k = 1.5$, at least 56 percent of cases lie between $(M_x - 1.5s)$ and $(M_x + 1.5s)$. (*b*) When $k = 4.0$, at least 94 percent of cases lie between $(M_x - 4s)$ and $(M_x + 4s)$.

standard deviation is 10. His deviation from the mean is 15, and the ratio of this deviation to the standard deviation is 1.5. Therefore this student is one and one-half standard deviations from the mean and from Tchebysheff's theorem we can assert that at least 56 percent of the students had scores closer to the mean (either higher or lower) than did this student. You will soon see that for the type of distribution usually encountered with test scores, *more* than 56 percent lie closer than 1.5 standard deviations from the mean.

3.4 Standard scores

The ratio of a deviation score to the standard deviation is probably the most useful transformation of a raw score into new units. This transformation is called a *standard score* or *z score*, and is symbolically defined as

$$z_{X_i} = \frac{x_i}{s_X}$$

$$= \frac{X_i - M_X}{s_X} \tag{3.4.1}$$

The advantage of z is that it is a pure number, free of the arbitrary original measurement units in which M_X and s_X are expressed. Thus scores from different distributions are more easily compared. The data presented in Table 2.8 can be used to illustrate the kind of comparison across distributions which z scores make possible. Suppose we are asked whether subject number 3 in Group 2, who is below average in his group, is more inferior in his group than is subject 3 in Group 1, who is also below average but who has a higher raw score. To answer the question, compute a z score for each subject:

$$z_{13} = \frac{(4-5)}{1.3} = -.76, \quad \text{Group 1, subject 3}$$

$$z_{23} = \frac{(3-5)}{3.0} = -.67, \quad \text{Group 2, subject 3}$$

The result indicates that although subject 3 in Group 1 has a higher raw score than does subject 3 in Group 2, he appears to be more deviant or inferior in his group (not to be said lightly in public of real people, of course).

Another way of defending the comparability of z score distributions arising from transformation of different raw score distributions is to observe that all z score distributions have the same basic distribution statistics M_z and s_z. In fact M_z is always zero and s_z is always 1. These constancies arise algebraically as follows:

$$M_z = \frac{\sum z}{N} = \frac{\sum (x/s)}{N} = \frac{(1/s) \sum x}{N}$$

but $\sum x$ is always zero, and therefore $M_z = 0$. Since M_z is zero, each z score is a deviation score, and from our definition of variance

$$s_z^2 = \frac{\sum z^2}{N}$$

Substituting x/s_x for z and taking the constant s_x out of the summation yields

$$s_z^2 = \frac{\sum (x/s)^2}{N} = \left(\frac{1}{s^2}\right)\left(\frac{\sum x^2}{N}\right)$$

Since $\dfrac{\sum x^2}{N}$ is s_x^2, this last expression reduces to:

$$s_z^2 = \frac{1}{s^2} \cdot s^2 = 1$$

Since the variance of z is 1, the standard deviation is also 1. This suggests that Tchebysheff's theorem can now be applied directly to z scores. For example, the absolute value of z ($|z|$, ignoring the algebraic sign) is greater than 1.5 no more than 44 percent of the time, $|z| > 2$ can occur only 25 percent of the time, and $|z| > 3$ is a rather rare event (never more than 11 percent of the z scores in the distribution).

Since z scores from different distributions reveal the relative positions of their owners in the different distributions, it is not surprising that educators favor the reporting of test and other assessment results in standard score terms. When a student's achievement report is presented as a profile of z scores, direct comparison of his relative position in the class on different measured abilities is possible. The spreading availability of electronic computers in school systems, which make feasible the routine computation of score transformations, may be expected to hasten the millenium when all educational evaluations will be reported in standard score fashion. Since such evaluations exist solely to allow comparisons, such a development is eminently desirable. A human problem which arises is that most people, even educators, don't like to deal with signed numbers, don't like to deal with decimal fractions, and abhor signed decimal fractions. The appearance of a z score such as -1.62 seems detestable to many. But even this aesthetic problem can be overcome, and in a way which illuminates the nature of the z transformation. Consider that Equation 3.4.1 could have been expressed as

$$z_{X_i} = \left(\frac{1}{s_x}\right) X_i - \frac{M_x}{s_x} \tag{3.4.2}$$

The astute student (or the one with a long memory) recognizes (3.4.2) as being in the form of an equation for a straight line, $y = ax + b$, where $a = 1/s_x$ and $b = M_x/s_x$. This means that the transformation from X to z is a *linear* (straight line) transformation of the raw scores. A linear transformation preserves the original ranking of the subjects and retains the shape of the frequency polygon. In order to overcome the mentioned shortcomings of z scores, they may be multiplied by a constant to clear the decimal fractions

STANDARD SCORES

and another constant may be added to clear the negatives. An attractive transformation is:

$$T_{X_i} = 10 z_{X_i} + 50 \qquad (3.4.3)$$

The resulting T scores will be whole, positive numbers in the range 0 to 100 (assuming the z scores were rounded to one decimal place and all scores are within 5 standard deviations from the mean).[2] These are easy scores to talk about, and are comparable from distribution to distribution. All such standard T score distributions will have mean 50 and standard deviation 10. (Multiplying all scores in a distribution by k multiplies the mean and the standard deviation by the absolute value of k, while adding k to a distribution adds k to the mean and leaves the standard deviation undisturbed. Why doesn't multiplying the z scores by 10 influence their mean?)

Adequate description of the variability of scores in a distribution is fully as important as description of the central tendency. Tables reporting means of distributions should also report standard deviations. Table 2.7 is modified as Table 3.2, and Table 2.6 is modified as Table 3.3, by the inclusion of standard deviations corresponding to all means. The new tables make us look at mean differences in the light of standard deviations. A later chapter will develop a complete strategy for analyzing between-group mean differences utilizing knowledge of within-group variabilities.

Table 3.2 Sophomore Test Battery Means and Standard Deviations for a Stratified Random Sample from All Public High Schools in New Hampshire[a]

Variable	College Preparation $N = 404$ Mean	Standard Deviation	Noncollege $N = 424$ Mean	Standard Deviation
SCAT Verbal	289	11	275	12
SCAT Quantitative	302	20	288	20
COOP Reading Vocabulary	156	7	148	8
COOP Reading Level	155	8	146	9
COOP Reading Speed	156	9	146	8
COOP Reading Expression	155	9	146	9

[a] From P. R. Lohnes and P. H. McIntire (1963).

[2] The authors use T as a handy acronym for "transformed score." However, T is sometimes reserved for outcomes of transformations that normalize the score distribution. The student should note that our T scores are simply linear transformations of z scores that will *not* do anything to normalize the distribution.

Table 3.3 Sophomore Test Battery Means and Standard Deviations for Two Curriculum Groups in Two Schools[a]

	College		Noncollege	
Variable	Mean	Standard Deviation	Mean	Standard Deviation
	School 1			
	$N = 239$		$N = 204$	
SCAT Verbal	293	12	276	11
SCAT Quantitative	305	24	293	13
COOP Reading Vocabulary	159	7	149	7
COOP Reading Level	156	8	146	8
COOP Reading Speed	159	8	149	8
COOP Reading Expression	158	8	146	8
	School 2			
	$N = 206$		$N = 156$	
SCAT Verbal	283	8	274	10
SCAT Quantitative	295	10	289	12
COOP Reading Vocabulary	154	5	149	7
COOP Reading Level	152	7	145	9
COOP Reading Speed	152	7	146	8
COOP Reading Expression	153	7	145	8

[a] From P. R. Lohnes and P. H. McIntire (1963).

Incidentally, the standard deviations reported in Tables 3.2 and 3.3 were not computed by Equation 3.3.1 exactly, but by a slightly modified variance formula, as follows:

$$\hat{s}_x^2 = \frac{\sum x_i^2}{N - 1} \qquad (3.4.4)$$

Note that the divisor for the sum of squares is $N - 1$ in (3.4.4), whereas it was N in (3.2.2). This new and slightly larger quantity, $\hat{s}_x^2$, is called the *unbiased estimate of the variance*. Most research publications report such unbiased estimates of variances and standard deviations based on them. Thus the standard deviations of Tables 3.2 and 3.3 were obtained from

$$\hat{s}_x = \sqrt{\frac{\sum x_i^2}{N - 1}} \qquad (3.4.5)$$

The need for $\hat{s}_x^2$ arises from consideration of the variation in variance estimates from sample to sample. That is, suppose we take a random sample of size N from a distribution of size N^1, a very large number, and that N is a small fraction of N^1. Then we compute s_x^2 for our sample by formula 3.2.2, and use it to estimate the variance for the population of all N^1 scores. This estimate may actually be larger than or smaller than the population variance for all N^1 scores. Do you see why this is so? But suppose we take many random samples, each of size N, from the population of N^1 scores, and compute s_x^2 for each sample, producing a sampling distribution for the variance. Some of these sample variances are larger than the population variance and others are smaller. Statisticians know that the average of these sample variances based on (3.2.2) will usually be slightly less than the population variance. The smaller divisor of (3.4.4) produces a slightly larger variance estimate from each sample, and insures that the average of the sample variances will tend to center itself over the population variance. This has been a confusing paragraph because it has opened a Pandora's box of new notions, such as sampling distribution and biased and unbiased estimates. The next several chapters concentrate on explaining these notions, and a set of computer demonstrations of them is coming in the laboratory portion of the course. They are fundamental notions of statistics, and later you will re-read this paragraph with a satisfying sense of insight. For the moment, please accept the dictum that although (3.2.2) defines s_x^2 as the variance of a sample of scores, the researcher usually computes as an unbiased estimate of the unknown variance $\hat{s}_x^2$, by formula 3.4.4, and an estimate of the standard deviation, $\hat{s}_x$, by (3.4.5), when he is processing sample data. Since behavioral science is the business of studying small samples of organisms in order to establish generalizations about entire populations of organisms, this dictum covers almost all research activity on behavioral problems. Statisticians use σ^2 (sigma square) as a symbol for the theoretical population variance which is estimated by s^2 or $\hat{s}^2$, and σ as a symbol for the population standard deviation. Thus, σ^2 and σ are population parameters, while s^2, $\hat{s}^2$, s and $\hat{s}$ are estimates, or statistics.

3.5 Summary

A statistic for the variability or scatter of scores around the mean in a distribution is provided by the *variance*:

$$s^2 = \frac{\sum x_i^2}{N}, \quad \text{where} \quad x_i = X_i - M_x$$

The variance is the average squared deviation from the mean. The *standard deviation* is the square root of the variance. Actually, researchers usually compute the *unbiased estimate of the variance* from their sample data:

$$\hat{s}^2 = \frac{\sum x_i^2}{N - 1}$$

Tchebysheff's theorem permits a statement of the proportion of the scores in any distribution which must lie within k standard deviations from either side of the mean ($k > 1$).

A very useful score transformation, which permits comparison of scores from different distributions, is

$$z_i = \frac{x_i}{s} = \frac{X_i - M}{s}$$

This linear equation defines the z score, or *standard score*, transformation. Standard scores are sometimes easier to deal with if they are converted to T scores:

$$T_i = 10z_i + 50$$

T scores have a mean of 50 and a standard deviation of 10. T and z scores will take on additional meaning after you have studied Chapter 7.

EXERCISES AND QUESTIONS

1. What is the basic difference between a statistic and a parameter? Give examples of each.

2. Compute the mean for the first ten cases in Table B.2 for variable 17 (Sociability). Then determine the deviation score (x) for each of the ten cases and compute the sum of them ($\sum x$). How does this compare with what you learned at the beginning of Section 3.2 about $\sum x$?

3. Find the sample variance of variable number 8 for the first twenty females of Appendix B. Use the procedure illustrated in Table 3.1. Although this is a convenient numerical exercise, what about the assumption regarding the scale properties of variable 8?

4. Obtain an approximate square root of the variance found in Question 3 and see if Tchebysheff's theorem holds for this set of twenty scores.

5. A student from the college curriculum group of School 1 in Table 3.3 had a score of 175 on the COOP Reading Speed Test. What would be his T score in his curriculum group's distribution?

6. Jane has a standard z score of +1.20. What is the minimum proportion of students that would fall between Jane's score and a score of −1.20 regardless of the shape of the frequency distribution?

7. Express Jane's standard z score as a T score.

8. Two first-grade teachers compare their new classes in the Fall and note that the means for IQ are almost identical and the means for Reading Readiness are very similar. They conclude that the two classes are so much alike that they can pool the work of planning for reading instruction this year, making one plan serve for both classes. What problems may arise from this decision? Why?

CHAPTER FOUR

Computing Statistical Analyses

4.1 Introduction to digital computers

We are living in a period in history in which a revolution is occurring. Since 1945 man's information processing and analyzing methods have been changing dramatically as a result of the invention of the electronic digital computer. You are aware, no doubt, of many of the implications of this development for the enterprises of science, business, industry, and government. The impact of the computer on statistics has been enormous. Statisticians have been freed of the drudgery of the computational chores that have always plagued them. Indeed, you as students are going to be spared arithmetic exercises that have plagued statistics students of the past. Furthermore, the results of a computer can be trusted for accuracy to a much greater degree than was ever possible with statistics computed by human operators. This is because of the large variety of checks that can be built into computer programs. The most important result, however, is that computers now enable statisticians to develop and apply methods that were impossible before. If you persevere in the study of statistics into advanced courses you will encounter examples of new styles in statistics the computer has made possible.

This introductory course is concerned with fundamental concepts and operations of "classical" statistics, but the computer is going to be used to run experiments that demonstrate the basic concepts. Through the agency of the computer you are going to see how statistical principles operate in data. Such understandings are difficult to obtain in courses taught without the assistance of the computer. You will be encouraged to learn the rudiments of computer programming in order that you may read and understand the computer programs that are run in the laboratory periods. You will *not* be required to write computer programs. All the programs to be used are provided. This is consistent with the emphasis in this course on having you acquire understanding and reading ability in statistics, rather than skills in doing specific statistical analyses. Many students who continue to study statistics in more advanced courses will want to become active program writers, but the emphasis here is for all students to get a general understanding of how scientists use computers.

The digital computer is a universal analytical machine. Turing's theorem (1936) proved that such a machine can be taught "to solve any problem that can be resolved into a finite sequence of symbol manipulations" (Feigenbaum and Feldman, 1963, p. 11). Machine intelligence is not a substitute for human intelligence, because there are successful human thought processes which

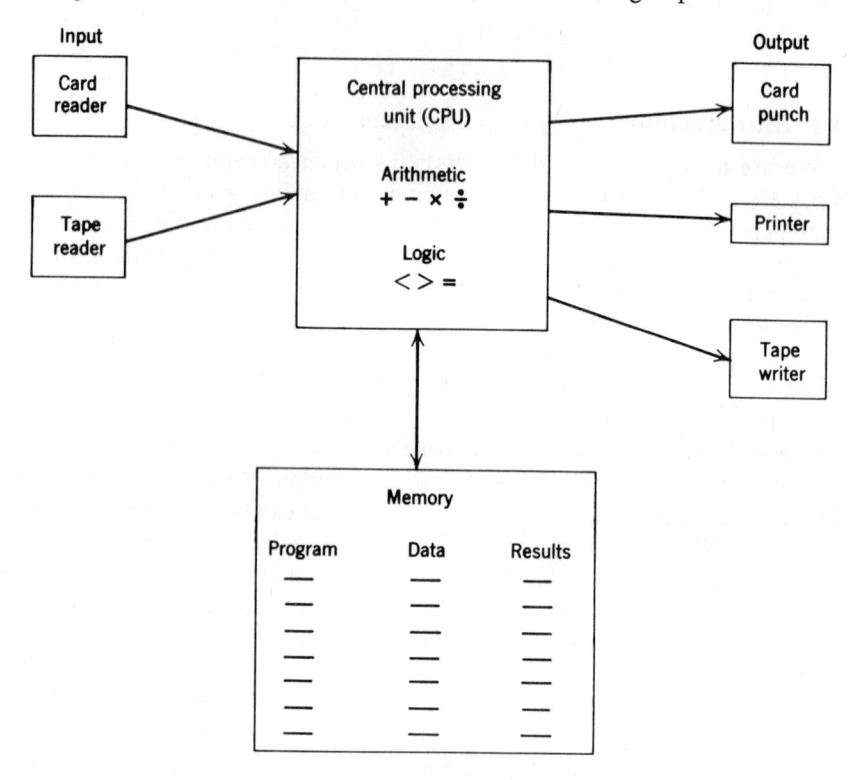

Fig. 4.1 Key components of the computer.

cannot be resolved into precise sets of rules. But the machine does duplicate many of man's most useful thinking strategies, and its repertoire is growing steadily. Also, there is no question that the computer has powers of thought denied man, because its astonishing speed of calculation, its complete persistence of and total access to its memory contents, and its accuracy by human standards, enable it to prosecute analytic strategies which man has invented but cannot employ.

The main components of a digital computer are illustrated in Figure 4.1. They include the following.

INTRODUCTION TO DIGITAL COMPUTERS 51

1. The *central processing unit*, often called "the main frame," which houses the electronic circuits, transistorized in the newer machines. These allow the machine to process information, and may be thought of as a central nervous system containing the built-in arithmetic and logical abilities.

2. The *memory*, usually called "core storage" in modern machines, which is capable of storing information in the form of binary (0 or 1) digits.

3. The *input–output devices*, which may be thought of as its sensory-motor equipment, through which it communicates with its environment. Among the great variety of input–output devices employed by computers are electric typewriters, card reading and punching units, and tape drives from which magnetic tape is read and written.

In your first visit to a computer center you will see that a computer is not just one big machine but a roomful of separate devices connected by cables running under the floor. Notice in Figure 4.1 that the memory device is used to store data, results, and the list of instructions or directions for going from data to results, called the "program."

Probably the outstanding feature of the digital computer is its capacity to operate by following a sequence of instructions that is stored inside the computer, in the memory. It is the stored program that makes the digital computer an example of Turing's universal machine. As a hardware system, a computer is stupid and ignorant in that it has inherently few abilities and little knowledge. As has been pointed out, it can learn any analytic ability or precise knowledge, because it has a memory that receives programs of instructions, which transfer abilities to it, and sets of data, which convey knowledge to it. Following an appropriate program, it can transform input data into new knowledge and report out its results. Most significantly, it can be taught rules for modifying its own program, so that the instructional sequences it obeys are systematically transformed. This last point is paramount, because there would be little value in having a machine capable of executing several hundred thousand instructions per second, if some human being had to prepare all those instructions. Most analytical processes involve the repetitive application of a small number of different operations (counting, comparing, changing, selecting, etc.) in extended sequences, with success depending on knowing when to terminate one sequence and initiate another one. The computer has the capacity to monitor such repetitive sequences and to make decisions about when to interrupt and replace them. Green (1963) has pointed out that "programming a digital computer is an exacting task, because the machine is such a relentless follower of instructions. It does just exactly what it is instructed to do, showing no common sense at all" (p. 14). As you will discover, you must be as precise in programming as the machine is relentless in following your program, but you will be tremendously relieved to discover how much of its own program the machine can create, following your guidelines.

People employ a fantastic conglomeration of languages, number systems, codes (musical notation, for example), gestures and grimaces in the nourishing and projecting of their cognitive processes. This semantic potpourri no doubt accounts for most of the glory and many of the agonies of human experience. Unlike humans, computers use a "two-letter" language. Literally everything is expressed in a two-digit or binary code (0 or 1). The writing of Shakespeare, when stored in a computer, is internalized as a long, long string of 0's and 1's. We would revert to our desk calculators, most of us, before we would learn to converse with a computer in its own binary language. Fortunately for us, computer scientists have written programs, called "compilers," which tell computers how to translate symbol systems preferred by human users into the binary tongue of the native machine. Among such compilers, one that has achieved almost universal acceptance among native American computers is called FORTRAN, for FORmula TRANslation language. FORTRAN closely approximates our everyday language and is easy to learn. For example, if we want the machine to read we write "READ," or to stop, we write "STOP." Different machines require slightly varying versions of FORTRAN, but the FORTRAN conventions presented in this chapter provide a fairly complete primer for most computers. These conventions are worth learning, even if you never write a program yourself, because university computer centers collect elaborate libraries of FORTRAN-coded statistical programs from which you may borrow as the need arises. You will understand and trust these borrowed programs only if you can read them. Sometimes you can make slight changes in them to make them suit your purposes better. It is easy to modify a borrowed program so that it prints *your* name and *your* project title at the top of the output, for example. Being able to read FORTRAN makes you an "insider" in the exciting world of computer users.

The next section is a primer on the features of the particular FORTRAN language used in the programs in this book. It will bear careful study now, but you should also expect to use it heavily as a reference as you think your way through the programs in this and later chapters.

The FORTRAN language has certain rules of grammar which must be followed precisely if the computer is to follow the FORTRAN programmer's directions. The primer which follows, then, is a set of rules which we must follow in writing FORTRAN statements. Before plunging into the details of these rules, it would be useful to get the "big picture" of what constitutes a typical program.

A FORTRAN program consists of a set of statements written in FORTRAN language, one statement per punched card. In general, the cards are arranged in the sequence in which the operations specified by the statements are to be performed by the computer. The exact arrangement of symbols in each card is specified toward the end of Section 4.2.

Typically, a program begins with a set of comment cards. These cards are not read by the computer as part of the program, but are explanations of the program for the human user. Comment cards can also be spread throughout the program to direct the user's attention to different sections of the program. They are identified as comment cards by a "C" punched in column 1.

Following the initial comment cards is generally a FORTRAN statement called a *DIMENSION* statement. *DIMENSION* statements are necessary to reserve large blocks of memory which may be needed to store subscripted variables. For example, the variable X_i, where subscript i goes from *1* to *N*, would require *N* storage locations in memory. If we just refer to the variable *X* without subscript, that refers simply to a single storage location in memory. In order to plan and reserve room for large arrays of subscripted variables, dimension statements provide the necessary information to the FORTRAN compiler.

The next part of the program generally consists of input statements. These are directions for the computer to read data from cards or some other input device. Input statements would also be used to read cards which specify the characteristics of a particular application of the program. For example, a computer program might be written to compute the means and standard deviations for up to fifty variables. A given user problem might consist of only twenty variables, so the first thing we have to tell the computer is how many variables are in the problem it is processing this time. Such information is read in at the beginning of the program under the direction of an appropriate FORTRAN input statement.

In the main body of the program, we might encounter a variety of FORTRAN statements. The most typical would be arithmetic statements which specify arithmetic operations to be performed on the data read into the computer under the direction of the input statements. For example, $A = B + C$ would be a possible FORTRAN arithmetic statement if the computer had previously read in numerical values for variables *B* and *C*. It would then add these two values and call the sum *A*. The sum *A* would then be available for subsequent processing or could be saved for a subsequent output statement. In addition to arithmetic statements, the main body of the program would also consist of certain types of control statements. These define conditions under which different operations would be performed. They also control the number of times a particular operation, such as addition, would be performed in the course of a program.

It is possible that not all of the operations to be performed within the program are completely specified within that particular program. For example, it is possible to refer the computer to a second program called a "subroutine." Once the operations defined in the subroutine are performed, the computer would then go back to the program which referred to the subroutine. An example would be the computations needed to perform square roots. Rather than writing the necessary arithmetic programming each time a square root is desired, a programmer could simply direct the computer to the square root subroutine every time it was required in the course of a program.

Once all the arithmetic is completed, a programmer then writes FORTRAN statements to allow him to see the results. These output statements may direct the computer to punch a card containing the computed results or print them out, depending upon the programmer's preference.

The final statement in a FORTRAN program is always *END*. This tells the compiler that it now has read all the FORTRAN statements for the present program and that the compilation should now be completed.

We suggest that you read through the remainder of Chapter 4, not with the idea of gaining mastery of all of the details regarding the rules of the FORTRAN language, but to get increased familiarity with the various aspects of the language. The details then will become clearer as you relate the programs at the end of the chapters to the FORTRAN rules presented in this chapter. The rules are introduced at this time so that you can begin to get familiar with them. The laboratory portion of the course, spread over the semester, will eventually allow you to develop mastery of them.

4.2 The FORTRAN language

Different types of variables and variable names

In FORTRAN there is a critical distinction between integers ("whole numbers") and numbers containing a decimal point. The programmer must maintain a strict observance of this distinction, and must identify the type of every number and every variable which takes numerical values. Integers are used in FORTRAN for counting and to identify the elements of a set. Variables that take integers for numerical values must be given names beginning with *I, J, K, L, M,* or *N*, and every variable with a name beginning with one of these six letters is an integer-type variable. Examples of integer-type variable names are *ID, ITEAM, N,* and *JTEAM*. Variables whose names begin with any of the other twenty letters of the alphabet are called floating-point variables. Numbers which contain a decimal point are called floating-point numbers because the computer employs a special notation for them, using an eight-digit number (mantissa) to preserve the eight most significant digits and a two-digit number (characteristic) to locate the decimal point. Examples of floating-point numbers are 8., .0, 100.37, and .00001. Examples of floating-point variable names are *ATEAM, X, Y,* and *RTEAM*. Variable

THE FORTRAN LANGUAGE 55

names may not contain more than six letters. Statistical variables are usually of the floating-point type. Variables can often be given meaningful names. Suppose we want to read into the computer a student's identification number, age, sex, SAT Verbal score, and SAT Quantitative score. A sensible FORTRAN statement would be

READ (5, 1) ID, AGE, SEX, SATV, SATQ

This makes *ID* an integer-type number, and all the others floating-point variables. The numbers in parentheses are explained later.

The FORTRAN compiler assigns a specific location in the computer memory to each variable name in the FORTRAN program. Thus *AGE* can be thought of as an address in memory where a specific number can be located. The *READ* statement establishes the location of the variables to be read. The *READ* statement would tell the computer to read a card and place the number for that person's age in a memory location to be called *AGE*.

Each such addressable location in memory is called a "word." Thus a computer word is the storage unit employed to represent a number in memory. Large computers store numbers in binary form, usually in a set of 36 binary digits.

Arithmetic statements

Arithmetic statements in FORTRAN are very similar to ordinary algebraic notation. The characters or symbols used for operators are

+ for addition
− for subtraction
* for multiplication
/ for division
** for exponentiation

Thus, $(A * B)/C$ means "multiply A times B and divide the product by C." $A ** 2.$ means "square A," and $A ** .5$ means "take the square root of A."

The = sign has a different meaning in FORTRAN than it does in regular algebra. It actually stands for the operation of replacement. $A = B$ means "put the numerical value of memory location B in memory location A, leaving the contents of B undisturbed." It is meaningful to write $A = A ** 2.$, for example, meaning "replace A with A-squared," for any value of A. In ordinary algebra, $A = A^2$ could only be true for $A = 0$ and $A = 1$, because = means equality rather than replacement.

Provision is made in FORTRAN for converting variables from one type to another. Thus $EN = N$ causes the computer to create a floating-point version of the integer N and place it in location EN. $N = EN$ causes an integer version of EN to be put in N. Any decimal fraction part of EN

would be dropped in such a conversion. These operations are called "floating" a variable and "fixing" a variable.

In FORTRAN arithmetic statements operations are done serially from left to right, with all exponentiation done first, followed by multiplications, divisions, and finally additions and subtractions. Parentheses are used to specify or change the order of events. For example, without parentheses, the FORTRAN statement

A = B * C + B/D − C ** .5

means in ordinary algebra

$$A = (B \cdot C) + (B/D) - \sqrt{C}$$

while a similar FORTRAN statement *with* parentheses

A = (B * ((C + B)/D) − C) ** .5

means algebraically

$$A = \sqrt{B \cdot \left(\frac{C + B}{D}\right) - C}$$

Function subroutines

The FORTRAN compiler provides the programmer with the right to call on a library of subroutines to accomplish certain frequently required transformations of a variable. Four frequently used library subroutines are:

ABS(X)	absolute value of X
SQRT(X)	square root of X
EXP(X)	e^X
ALOG(X)	log e of X

A typical use would be the statement

A = ABS(A) * LOG(B)

This one FORTRAN statement would tell the computer to take the absolute value of variable *A* (disregard any negative sign), multiply that by the natural logarithm of variable *B*, and store the product in the memory location called *A*. It is possible to write additional function subroutines as they are needed. Examples are given in Chapter 5.

Control statements

Any FORTRAN statement may be numbered to allow a branching of control from another part of the program to it. The statement number precedes the statement in the punched card in columns 2 to 5. The easiest

THE FORTRAN LANGUAGE

control statement is the *GO TO* statement. It accomplishes an unconditional branching. Thus

$$\text{GO TO } 12$$

instructs the computer to branch to FORTRAN statement number *12*. The most useful control statement is the *IF* statement. It has the general form

IF (X) 2, 3, 4

which causes the computer to test the numerical value of *X*, and if it is negative to branch to statement *2*, if it is zero to branch to statement *3*, and if it is positive to branch to statement *4*. Thus

IF (K/2 − L) 33, 41, 7

means

$$\text{If } [(K/2) - L] < 0, \quad \text{go to 33}$$
$$\text{If } [(K/2) - L] = 0, \quad \text{go to 41}$$
$$\text{If } [(K/2) - L] > 0, \quad \text{go to } 7$$

A *STOP* statement terminates execution of a program.

Subscripted variables

The algebraic notation

$$X_i, i = 1, 2, \ldots, n$$

is indispensable for notating a variable which is actually an array of numbers in this case *n* of them. Since the machine for punching cards is not capable of dropping a character half a space to indicate a subscript, as the typewriter can, FORTRAN subscripts are placed in parentheses, and a subscripted variable appears as *X(I)*.

In translating FORTRAN statements into a machine-language program, the FORTRAN compiler must assign memory space to every variable. A variable which can have only one value at a time is no problem because it obviously requires just one address in memory. A subscripted variable is a problem for the compiler because it is an array of numbers, each requiring its own address. That is, one location of memory is assigned to each value of the subscript. To make matters worse, the range of the subscript, *n* for example, may itself be a variable. The solution to this problem is for the programmer to specify the maximum size of *n* of any use of the program and for the compiler to set aside that many memory cells for the subscripted variable. Then if all the designated cells are not used, no harm is done. The designation of the maximum length of an array is accomplished by a *DIMENSION* statement. Every subscripted variable used in a program must be dimensioned before it is used. An example is

DIMENSION   X(100), Y(100), TEAM(10)

which reserves 100 memory positions for $X(1), X(2), \ldots, X(100)$, another 100 positions for $Y(1), Y(2), \ldots, Y(100)$, and 10 positions for $TEAM(1)$, $TEAM(2), \ldots, TEAM(10)$. All these positions need not be used in the execution of the program, but of course the subscripts on X and Y must never exceed *100*, and on *TEAM* must never exceed *10*.

DO *loops*

The computer's ability to modify its own instructions systematically is the most significant advantage of stored programs, as it permits a few instructions to be executed a great many times in the performance of a repetitive task. A sequence of instructions which is executed repeatedly is called a "loop." The *IF* statement may be the basis for a decision to repeat a loop or to exit from it. For example, consider the sequence

```
      N = 1000
      SUM = 0.0
      J = 1
    1 SUM = SUM + X(J)
      J = J + 1
      IF (N − J) 2, 1, 1
    2 CONTINUE
```

This sequence computes the sum of 1000 items in the array $X(J)$. When the value of J reaches 1001 it is larger than N and the *IF* statement terminates the loop and branches control to statement 2. Such sequences of repetitive operations occur so frequently in statistical programs that we can be very thankful for the provision in FORTRAN of an easier way of controlling them. The easier way is by using a *DO* statement, of the form

```
      DO 5  I = 1, N
```

where 5 is a statement number for a statement somewhere ahead of this one in the program. This *DO* statement sets up a loop to execute all statements ahead, through statement 5, N times, setting $I = 1$ the first time, and incrementing I by *1* each time, so that on the last time through the loop $I = N$. The loop controlled by the *IF* statement in the example above would look like this, if controlled by a *DO* statement:

```
      N = 1000
      SUM = 0.0
      DO 1  J = 1, N
    1 SUM = SUM + X(J)
      CONTINUE
```

THE FORTRAN LANGUAGE

By the way, *CONTINUE* is an acceptable nonexecutable statement in FORTRAN, used as the first instruction after the loop in these examples because we don't know what the next operation might be. This statement is actually needed in a FORTRAN program to end a loop which otherwise would end on an *IF* statement, since it is illegal to place an *IF* statement as the last statement within a *DO* loop. Raising the possibility of an *IF* statement occurring within a *DO* loop suggests that it is possible to branch ahead within a *DO* loop, or to branch out of a *DO* loop. It is not possible to branch into a *DO* loop from outside it. *DO* loops may also be nested; that is, one *DO* loop may start and finish within another *DO* loop. These may be confusing points, but close scrutiny of the examples to follow will clarify the correct uses of *DO* loops.

Input-output statements

The most finicky part of FORTRAN programming is the business of getting information into and out of the computer. Yet the biggest advantage of FORTRAN lies in the *relative ease* with which input and output statements are coded, as compared with the difficulties of machine-language coding in this area. In FORTRAN the computer is told to read an IBM card by a *READ* statement and to punch an IBM card by a *WRITE* statement. In each case a "list" of the contents of the card follows the instructional word. Thus we have already had the example

READ (5, 1)   ID, AGE, SEX, SATV, SATQ

in which the five variable names are the list. The difficulty comes in specifying to the computer exactly how the variables are punched on the IBM card (or are to be printed or punched, in the case of a *WRITE* statement). The word *READ* or *WRITE* is always followed by two integers in parenthesis, before the list begins. The first integer tells what physical device is to be used (for example, the card reader) for reading or writing. For reading a card the number is generally 5. In a *WRITE* statement, the number 6 is the convention used here for printing a line and 7 indicates punching a card. The second integer is a statement number and refers to a *FORMAT* statement which contains the specification of the punched card arrangement. The *FORMAT* statement for the *READ* list in the example above might be

1   FORMAT (I5, F3.0, F1.0, 6X, 2F5.0)

In order to unscramble this, you must realize the following things.

1. Punched cards contain eighty columns in which digits or other characters may be punched. The set of columns occupied by each variable punched on a card is called its "field."

2. The *FORMAT* statement has to specify the fields for the variables in order of their appearance on the card, from left to right. The "field width," or number of columns, for each variable must be specified.

3. Every column from the first to the rightmost column used on the card must be accounted for in the *FORMAT*. Unused columns between variables may be accounted for by including blank columns in the field widths of variables, or by specifying an X-field, which is a "skip" field, telling the computer to ignore the contents of a number of columns.

4. Each variable must be identified as to its type. The codes used to type variables in *FORMAT* statements include

    I   Integer type variable
    F   Floating-point type variable
    E   Floating-point variable punched in exponential notation
    X   A skip field
    H   Hollerith field to be stored for labeling output

A Hollerith field is one containing any mix of the digits, alphabetic letters, and other characters which are employed in FORTRAN. The content of an *H*-field is actually punched in the FORMAT card in the x columns following the specification xH, and is reproduced in the same position in any card punched by a *WRITE* statement which references this *FORMAT*, for labeling purposes.

If x is an integer specifying a field width, and y is an integer specifying the number of adjacent fields described, when the field description applies to two or more variables in a row on the card, and z specifies the number of decimal places to the right of the decimal point, the placement of x, y, and z are

$$y \, I \, x$$
$$y \, F \, x \cdot z$$
$$y \, E \, x \cdot z$$
$$x \, X$$
$$x \, H$$

Field specifications are separated by commas, except that no comma is needed after the specification of an *H*-field.

Now we can unscramble our example which looked like this:

    READ (5, 1)   ID, AGE, SEX, SATV, SATQ
1   FORMAT   (I5, F3.0, F1.0, 6X, 2F5.0)

The computer is promised a data card on which columns 1–5 contain an integer-type identification number, columns 6–8 contain an age code to be

THE FORTRAN LANGUAGE

read as a floating-point variable (*F*-fields may or may not contain an actual decimal punch, at the user's option, in data cards), column 9 contains a sex code to be read as a floating-point variable (obviously it has been punched without a decimal point), columns 10–15 are to be ignored, columns 16–20 contain a SATV score, and columns 21–25 contain a SATQ score. Notice that it is not necessary to indicate that the remaining columns (26–80) are to be skipped. Figure 4.2 pictures an actual score card punched according to this *FORMAT*. Note that no decimal points have been punched, and that the SATV and SATQ fields contain two blank columns. Can you see that the *FORMAT* specifies all the floating-point variables as whole numbers, without fractional parts? *FORMAT* statements for *WRITE* statements (as opposed to *READ* statements) must always include a column for a decimal point and

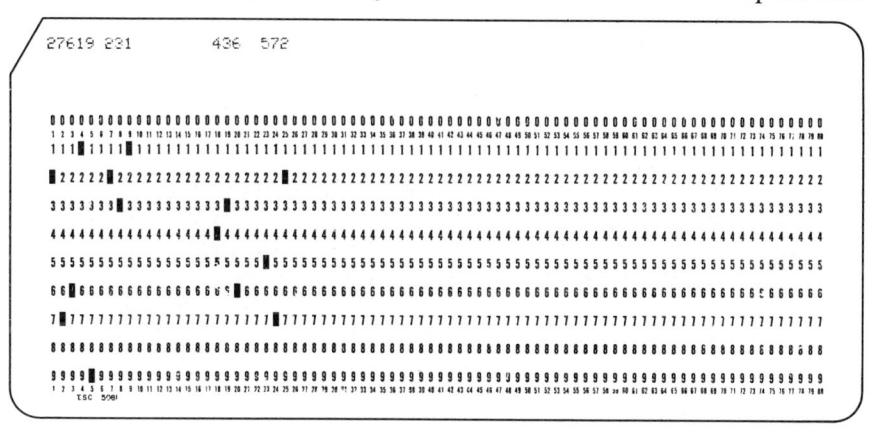

Fig. 4.2 Score card for Student 27619.

a column for a sign in the field width of any *F*-type variable, and a column for a sign in the field width of any *I*-type variable. Also on *WRITE* statements, the first character in each line controls the spacing on the printer, so that if no special spacing is required for that line, the first character is left blank. If the first character is a "1," the printer spaces to a new page before printing that line.

As an example of the use of *H*-fields,[1] suppose you are Sally Jones and you have written a program to sum *N* numbers. You would like to identify your output appropriately, so you include these statements:

```
    WRITE (6, 10) N
10  FORMAT (7H SUM OF I4, 18H NUMBERS. S. JONES)
```

[1] Some of the newer FORTRAN compilers allow Hollerith text placed between apostrophes. This eliminates the need to count out each character.

If $N = 100$ on a particular execution of the program, a line of output is written which contains the following

SUM OF 100 NUMBERS. S. JONES

When reading or writing an array of numbers it is necessary to supply the loop control and the incrementing subscript. FORTRAN includes a feature called "list control" for this purpose. For example, to read N elements into $X(J)$:

    READ (5, 10)   (X(J), J = 1, N)
10   FORMAT (20 F4.0)

This format would expect 20 values of X per card. If N is less than or equal to 20, only one card would be read. To punch the N elements of $X(J)$, four per card, each preceded by its position number, J:

    WRITE (7, 20)   (J, X(J), J = 1, N)
20   FORMAT (4(3X, I5, 2X, F5.0))

If $N = 16$, then four cards would be punched. Figure 4.3 is an example of the input card read according to format statement *10* and the first output card punched by format statement *20*. Notice the decimal point was punched even though no decimal places were indicated.

Compiler instructions and a program example

Two types of cards are placed in FORTRAN programs which have nothing to do with the design of the program. One is a "comment" card, which contains notes on the program for the benefit of the human reader. Such cards are identified by a C punched in the first column, which instructs the compiler to ignore the card. (No other character ever appears in the first column of a FORTRAN card. Columns 2–5 are used for statement numbers, and columns 7–72 are used for statements.) The other is the *END* card which is the last card in every FORTRAN program deck. It tells the compiler that all the program statements have been read.

To illustrate some of the FORTRAN features, let us build a little program capable of reading a set of N numbers, computing their sum, and reporting the sum. We always start with a comment card that identifies the program:

   C SUM OF N NUMBERS

Then we arrange to read in N (the number of numbers to be summed), which was punched according to format statement number 1.

    READ (5, 1) N
1   FORMAT (I5)

THE FORTRAN LANGUAGE 63

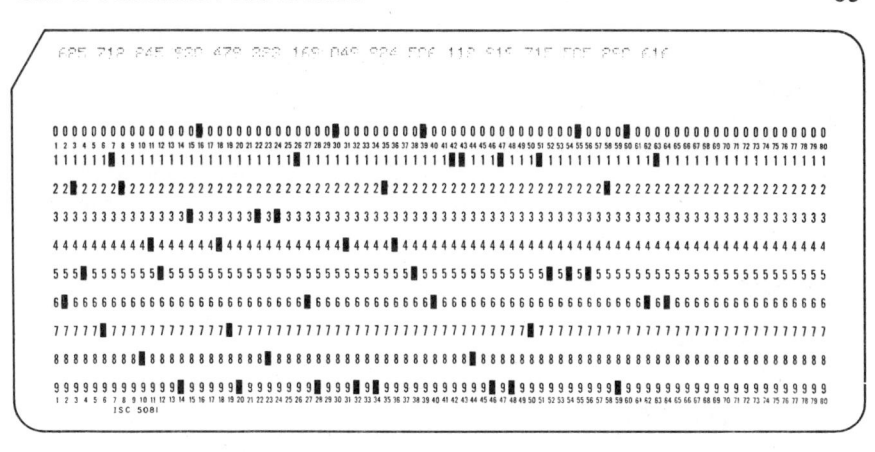

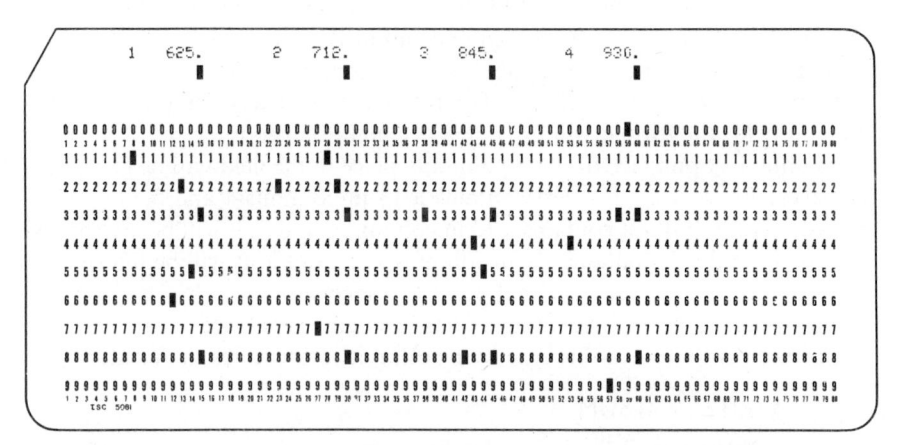

Fig. 4.3 Corresponding input and output cards.

Note that the current value of N is to be punched in the first five columns of a data card. N must be "right-adjusted" or put all the way to the right in the five-column field, because blank columns are read as zeros in data cards. That is, if N is 45, and it is punched in columns 1 and 2, the computer will read N as 45000. The format calls for 45 to be in columns 4 and 5.

Have you figured out what the computer does with number 45, which it reads from the first data card? When the program is compiled, a storage location in memory is given the name N, and the execution of the instruction to $READ\ (5,\ 1)\ N$ causes the number found in columns 1–5 of the data card, say 45, to be stored in the location named N.

The next task is to initialize a memory location in which we are going to accumulate the sum:

SUM = 0.0

This is like clearing the accumulator of an adding machine before starting to add.

Now we set up a loop with the range *1* through *N*, in which the computer will read each number and add it to *SUM*:

    DO 2 J = 1, N
    READ (5, 3) X
2 SUM = SUM + X
3 FORMAT (F5.0)

The *DO* statement establishes that all the statements down through statement *2* will be performed repetitively *N* times. When *J* reaches the value *N* the 45 numbers will have been read and summed. Note that the format for a number specifies that each number will occupy columns 1 through 5 of a separate data card. If no decimal point is punched anywhere in the five-column field, the format will provide a decimal point to the right of the digit in column 5. A punched decimal point any place in the field will be respected, however. Thus, if a data card contains a 13 in columns 4 and 5, it is interpreted as 13.0, but if it contains a 13 in columns 2 and 3, it will be interpreted as 1300.0, and if it contains 13. in columns 1, 2, and 3, it will be interpreted as 13., since the punched decimal point is respected.

Now that the sum of the numbers has been computed, we need to bring it out of the computer, in this case by printing it (*6* identifies the printer).

    WRITE (6, 4) SUM
4 FORMAT (F15.4)

The output format provides for a large field definition, since we don't know what the magnitude of the sum is likely to be. We could output the sum on a punched card if we wanted to (*7* identifies the card punch).

    WRITE (7, 4) SUM

Next, we terminate the program:

    STOP
    END

The *END* in columns 7–9 tells the FORTRAN compiler that this is the end of the program to be compiled, thus starting compilation.

THE FORTRAN LANGUAGE

Our complete program appears as follows, punched in the indicated columns.

Column	1	2	3	4	5	6	7--------------72
	C						SUM OF N NUMBERS
							READ (5, 1) N
		1					FORMAT (I5)
							SUM = 0.0
							DO 2 J = 1, N
							READ (5, 3) X
		2					SUM = SUM + X
		3					FORMAT (F5.0)
							WRITE (6, 4) SUM
		4					FORMAT (F15.4)
							WRITE (7, 4) SUM
							STOP
							END

Usually, there are several ways to do a particular job in FORTRAN and many ways to improve the output. Let's consider some variations on our summing program. Suppose we have our numbers prepunched 40 to a card, in two-column fields. How can we read 800 numbers, say, from 20 such cards? After reading N as 800, we might read all 800 numbers into an array called $X(J)$. To make this possible, we would start the program with a *DIMENSION* statement:

    DIMENSION X (1000)
C N IS NOT TO EXCEED 1000

Now let's label the printout:

    WRITE (6, 5) N
5 FORMAT (10H1 SUM OF I6, 8H NUMBERS)

This will cause the printer carriage to start printing on a new page (due to the *1* after the first *H*, which is a carriage control code) and to print the line:

    SUM OF 800 NUMBERS

Next, we read in the 800 numbers under list control:

    READ (5, 3) (X(J), J = 1, N)
3 FORMAT (40F2.0)

Can you see that the format provides for getting 40 two-digit numbers from each card? Next, we modify our *DO* loop to recognize that the numbers are already stored in $X(J)$:

```
    DO 2  J = 1, N
  2 SUM = SUM + X(J)
```

Now, let's label the sum explicitly:

```
    WRITE (6, 4) SUM
  4 FORMAT (7H0SUM = F15.4)
```

The 0 after the *H* is a carriage control that causes a double space before printing. If the sum of the 800 numbers is 28,753, the output line will be:

```
SUM = 28573.0000
```

Finally, if we make the first operable statement number *10*:

```
 10 READ (5, 1) N
```

and make the last operable statement *GO TO 10* rather than *STOP*, when the first summation problem is finished the computer will go back to the card reader and seek another. In this way, a whole set of summing problems may be executed, one right after another. When the computer finishes the last problem, it will sit idle awaiting the pleasure of the operator.

The modified program looks like this:

```
    C    SUM OF N NUMBERS.
         DIMENSION X(1000)
      10 READ (5, 1) N
       1 FORMAT (I5)
         WRITE (6, 5) N
       5 FORMAT (10H1 SUM OF I6, 8H NUMBERS)
         READ (5, 3)(X(J), J = 1, N)
       3 FORMAT (40F2.0)
         SUM = 0.0
         DO 2 J = 1, N
       2 SUM = SUM + X(J)
         WRITE (6, 4) SUM
       4 FORMAT (7H0SUM = F15.4)
         GO TO 10
         END
```

THE FORTRAN LANGUAGE

Note that the statement numbers are mere labels and do not have to appear in numerical sequence. Also, FORMAT statements can be placed anywhere in the program since they do not specify operations. They only provide information regarding data arrangements and must always be numbered for reference purposes.

As another example, consider the following program for creating a table of squares and square roots for integers from 1 to 1000.

```
C   SQUARES AND SQUARE ROOTS OF INTEGERS
    WRITE (6, 5)
5   FORMAT (1H1)
    J = 0
1   J = J + 1
    XJ = J
    XJSQ = XJ * XJ
    XJRT = SQRT (XJ)
    WRITE (6, 2) J, XJSQ, XJRT
2   FORMAT (1H0, I6, 11H, SQUARE = F10.0, I6H,
    SQUARE ROOT = F10.6)
    IF (1000 − J) 3, 3, 1
3   STOP
    END
```

Note that the effect of *WRITE* (6, 5) is simply to cause the printing of the table to start on a fresh page. The *IF* statement was used to illustrate how a loop may be controlled by it, but you should be able to see that a *DO* loop would simplify the program. Write the same program using a *DO* loop instead of the *IF*, as a test of your understanding.

This has been the barest primer on elementary features of FORTRAN, and we realize you need assistance in your efforts to learn the features of this new language. Mostly we learn a language by using it, and we urge you to practice FORTRAN by carefully studying the programs in this book. If you are not getting some assistance in a laboratory section of this course, you may want to study FORTRAN more intensively on your own. IBM has a self-instructional course on basic FORTRAN that you can obtain.[1] Alluisi (1967) has prepared an excellent paperback guide titled, *Basic Fortran for Statistical Analysis* (Dorsey Press, 1967).

4.3 Computing distribution statistics

This section describes and lists programs for computing means and variances, and transforming raw scores to deviation, z, and T scores. The student should punch up these programs and run them with appropriate Project TALENT data in Appendix B. Close attention to the features of these programs will be rewarded repeatedly, as the methods of reducing score rosters to accumulations (sums and sums of squares) are basic to all statistical operations. The chapter concludes with a listing for a program which computes a frequency table for a score distribution. We also give a subroutine version of this distribution program. The student should run this program with the TALENT data and convert the resulting tabulation into a frequency polygon or histogram.

Computing the mean

There are several ways to program the computation of a mean in the FORTRAN language. The two methods to be demonstrated differ in that in the first, one score after another is read into a place in memory called X, added to SUM, and replaced, so that only one score is present in the computer at any stage in the process. In the second method, all the scores are read into a vector, $X(J)$, then one score at a time is added to SUM. To illustrate that this second method retains all the scores in the vector, or subscripted variable, $X(J)$, after the mean is computed it is subtracted from each score and the resulting deviation scores are placed in another vector, $Y(J)$. Both programs require initialization by reading in a value for N—the number of scores to be averaged—and setting the accumulation variable, SUM, equal to zero before accumulation begins. Note the two ways of controlling the summing loop employed in the programs. Which is superior? Why is the statement $EN = N$ included early in each program? How are the data to be punched for program *MEAN*? For program *DEVSCOR*? (*Hint:* MEAN reads one score per card, while DEVSCOR reads twenty scores from each card.) Note the list control "implied *DO* loop" controlling the reading of scores in *DEVSCOR*, and contrast it with the explicit *DO* loop by which the resulting deviation scores are punched. Each program will operate on any number of jobs without restarting and will terminate when all the data are processed.

[1] FORTRAN IV for System/360 Programmed Instruction Course, Forms R29-0080 through R29-0087, IBM DPD, 112 East Post Road, White Plains, New York 10601, or call your local IBM office.

```
C      PROGRAM MEAN
C      COMPUTES AND REPORTS THE SUM AND MEAN OF
         N SCORES.
    1  READ (5, 2) JOBNO, N
    2  FORMAT (2I5)
       WRITE (7, 3) JOBNO, N
    3  FORMAT (9H JOB NO. I6, 10H, MEAN OF I6, 8H SCORES.)
       EN = N
       SUM = 0.0
    4  READ (5, 5) X
    5  FORMAT (F5.0)
       SUM = SUM + X
       N = N − 1
       IF (N) 6, 6, 4
    6  XM = SUM/EN
       WRITE (7, 7) SUM, XM
    7  FORMAT (7H SUM = F12.4, 10H , MEAN = F12.4)
       GO TO 1
       END
```

```
C       PROGRAM DEVSCOR
C       COMPUTES MEAN AND DEVIATIONS FOR A SET OF N
        SCORES.
        DIMENSION X(1000), Y(1000)
1       READ (5, 2) JOBNO, N
2       FORMAT (2I5)
        WRITE (7, 3) JOBNO, N
3       FORMAT (9H JOB NO. I6, 24H, PROGRAM DEVSCOR,
        N = I6)
        EN = N
        SUM = 0.0
        READ (5, 4) (X(J), J = 1, N)
4       FORMAT (20F4.0)
        DO 5   J = 1, N
5       SUM = SUM + X(J)
        XM = SUM/EN
        WRITE (7, 6) SUM, XM
6       FORMAT (9H SUM = F9.4, 10H, MEAN = F9.4)
        DO 7   J = 1, N
        Y(J) = X(J) − XM
7       WRITE (7, 8) J, X(J), Y(J)
8       FORMAT (9H SUBJECT I6, 13H RAW SCORE = F5.0,
        6HDEV = F9.3)
        GO TO 1
        END
```

Computing the variance

The variance of a set of numbers is easily computed by accumulating the sum of squared scores in the same loop in which the sum of scores is accumulated. First, initialize an accumulator:

SXSQ = 0.0

Then insert within the summing loop either the statement

SXSQ = SXSQ + X * X

or, if the scores have been read into a dimensioned array

SXSQ = SXSQ + X(J) * X(J)

COMPUTING DISTRIBUTION STATISTICS

After the accumulations loop is completed, the variance is computed from the raw sum and the raw sum of squares by application of Equation 3.4.4:

```
    SSD = SXSQ - SUM * SUM/EN
    VARX = SSD/(EN - 1.0)
    SDX = SQRT(VARX)
    WRITE (7, 117) VARX, SDX
117 FORMAT (12H VARIANCE = F8.2, 9H, S.D. = F8.2)
```

If these computations are inserted into program *DEVSCOR*, they provide the standard deviation, *SDX*, needed to convert deviation scores to standard scores. This may be accomplished by inserting the following loop. Can you figure out where to put it?

```
    DO 8  J = 1, N
    Y(J) = Y(J)/SDX
  8 WRITE (7, 108) J, X(J), Y(J)
108 FORMAT (9H SUBJECT I3, 13H RAW SCORE = F5.0,
    4HZ = F8.3)
```

Note that this loop assumes $Y(J)$ already contains the deviation score for the *J*th person. Does it? As a further step in transforming *X*, *T*-scores may be computed:

```
    DO 9  J = 1, N
    Y(J) = 10.0 * Y(J) + 50.0
  9 WRITE (7, 109) J, X(J), Y(J)
109 FORMAT (9H SUBJECT I3, 13H RAW SCORE = F5.0,
    3HT = F8.3)
```

FORTRAN LIST 4.1 combines these additions to DEVSCOR into one coherent program, and provides a test problem and the output from the test problem.

FORTRAN LIST 4.2 gives you a program for producing frequency and cumulative frequency tables for grouped or ungrouped data. It can also rank up to 999 observations, if a ranking is desired. It, too, comes with test problem input and output. It is essential that you get this program running, as it will play an important role in the laboratory work for the rest of the course. We hope that you will study this program carefully and puzzle out its methods of building up the frequency and cumulative frequency distributions and ranking the data. This little program is going to spare you a great deal of odious labor this semester once you have learned how to use it.

You will find that it is necessary to provide a small set of "job control" cards ahead of your FORTRAN program when you submit your job to the computer in your computer center. These cards describe the job to the computer. They tell the computer who the job is to be billed to, that FORTRAN is to be compiled, that data are to be read from cards, and that the compiled program is to be executed. We are not able to specify these cards for you because they differ from center to center, even for a given computer. Every computer center has a professional staff able and willing to show users what control cards particular types of jobs require. All the programs in our series will use the same set of control cards, so you will need to seek advice only once.

4.4 Summary

The digital computer is noteworthy for its inhuman speed and accuracy, but its main feature is its stored program, which allows it to incorporate into its memory a sequence of instructions prepared by the user. FORTRAN is the name of the coding language employed by most statisticians. It is a simple language, but is very powerful, particularly because of the *DO* loop for running the subscripts of subscripted variables.

FORTRAN LIST 4.1 provides a program for computing means and standard deviations, and converting raw scores to deviation scores, z scores, and T scores.

FORTRAN LIST 4.2 provides a program for computing means and standard deviations, frequency distributions, cumulative frequency distributions, and (optionally) ranking the scores.

FORTRAN LIST 4.3 provides a subroutine for computing distributions.

EXERCISES AND QUESTIONS

1. What would *SUM* equal following these four statements?
    ```
    J = 1
    SUM = 19.5
    X(J) = 3.2
    SUM = SUM + X(J)
    ```

2. What would *A* equal after the following FORTRAN sequence was completed?
    ```
          A = 2.0
          DO 12  I = 1, 8
          B = I
       12 A = A + B
    ```

SUMMARY

3. What value does J equal following these three steps?

    K3 = 4
    M2 = 11 * K3
    J = M2/20

4. Write the original formula for the following FORTRAN statement.

    R = (A + B * X)/(C + D * X)

5. What statement number will the computer go to after the *IF* statement?

    M = 5
    K = 2 * 3 − M
    IF (K) 2, 100, 7

6. Are the following valid or invalid arithmetic statements?

    (a) JACK = JOHN − FRANK
    (b) X = T + 3 * L
    (c) C = X ** 3 + A

7. Write a *FORMAT* statement that would tell the computer how to read the student *ID*, and his scores for variables number 6 (weight) and 17 (sociability) in the Appendix B cards.

8. Using the abstract reasoning score for the males in Appendix B.2 (variable 15), run the program List 4.1. Then run the List 4.2 program on the four different scores punched out for each student in List 4.1 (X, x, z, and T). A comparison of the four distributions and four sets of statistics produced for these four variables by List 4.2 should be a good review of Chapters 2 and 3. We suggest you use eight class intervals with *EL* set equal to ($M_X - 4S$) and $EI = 5$.

```
C     LIST 4.1    TRANSFORMED SCORES
C
C          THIS PROGRAM COMPUTES THE MEAN AND STANDARD DEVIATION OF   N   SCORES
C     (MAXIMUM N IS 400) AND REPORTS THE DEVIATION SCORE, STANDARD SCORE, AND
C     T-SCORE TRANSFORMATIONS FOR EACH SCORE.  THE FIRST DATA CARD SPECIFIES
C     (COLS 1-5)  NPROB,   THE PROBLEM NUMBER
C     (COLS 8-10)  N,  THE NUMBER OF OBSERVATIONS
C     AND IS FOLLOWED BY N CARDS, EACH CONTAINING AN I.D. NUMBER (COLS 1-5) AND
C     A SCORE (COLS 6-10).
C
C
      DIMENSION   ID(400),   X(400),   DEVX(400),   Z(400),   T(400)
C
   1  READ(5,2)    NPROB,  N
   2  FORMAT (2I5)
      EN = N
      SUM = 0.0
      SXSQ = 0.0
      DO 10    J = 1, N
  10  READ(5,3) ID(J),   X(J)
   3  FORMAT (I4,22X,F3.0)
C     THIS IS THE MODIFIABLE FORMAT FOR INPUT SCORES.
C
      DO 4   J = 1, N
      SUM = SUM + X(J)
   4  SXSQ = SXSQ + X(J) * X(J)
C     STATEMENT 4 ENDS THE ACCUMULATIONS (SUMS AND SUMS OF SQUARES) LOOP.
      XM = SUM / EN
      VARX = (SXSQ - SUM * SUM / EN) / (EN - 1.0)
      SDX = SQRT  (VARX)
      WRITE(6,5)         NPROB,   N
   5  FORMAT (30H1TRANSFORMED SCORES.  PROB. NO. I6, 6H.  N = I6)
      WRITE(6,6)    XM,   SDX,   VARX
   6  FORMAT (8H0MEAN = F10.5,10H    S.D. = F10.5, 13H  VARIANCE = F10.5)
      DO 7   J = 1, N
C
      DEVX(J) = X(J) - XM
      Z(J) = DEVX(J) / SDX
   7  T(J) = 10.0 * Z(J) + 50.0
      WRITE(6,8)
   8  FORMAT(42H0   I.D.       SCORE       DEV        Z        T)
      DO50    J = 1,N
  50  WRITE(6,9)    ID(J),X(J),  DEVX(J), Z(J),  T(J)
   9     FORMAT (I6,  4X,  F6.0,  4X,  F6.1,  4X,  F6.3,  4X,  F6.2)
      CALL EXIT
      END
```

```
C     LIST 4.2   UNIVARIATE DISTRIBUTION
C
C           THIS PROGRAM PRODUCES THE MEAN AND STANDARD DEVIATION, A FREQUENCY
C     DISTRIBUTION, AND A CUMULATIVE FREQUENCY DISTRIBUTION FOR A SINGLE
C     VARIABLE.
C
C     THE FIRST CARD IN THE INPUT DECK CONTAINS
C     (COLS 1-5)   NPROB, THE PROBLEM NUMBER
C     (COLS 6-10)  N, THE NUMBER OF OBSERVATIONS TO BE DISTRIBUTED
C     (COLS 11-15)   K, THE NUMBER OF CLASS INTERVALS
C     (COL 20)   NRANK, A CONTROL VALUE THAT IS SET TO ZERO (OR LEFT BLANK),
C          UNLESS THE USER DESIRES TO HAVE THE PROGRAM RANK THE OBSERVATIONS AND
C          PRINT OUT A LARGEST-TO-SMALLEST RANKING.  NRANK = 1  CAUSES RANKING,
C          BUT NOTE THAT THE PROGRAM AS DIMENSIONED CANNOT RANK MORE THAN 200
C          OBSERVATIONS.
C
C                  THE SECOND INPUT CARD CONTAINS LOWER LIMITS FOR K CLASSES,
C     IN FIVE COLUMN FIELDS (INCLUDE A DECIMAL POINT IN EACH FIELD).  IF MORE THAN
C     16 CLASSES ARE TO BE ESTABLISHED, A SECOND CARD MAY BE USED TO SPECIFY
C     LOWER LIMITS (AS MANY CARDS AS REQUIRED, AS LONG AS EACH CARD DEFINES
C     16 CLASSES).
C
C     THE NEXT CARDS ARE SCORE CARDS.
C                  NOTE THAT EACH OBSERVATION IS ON A SEPARATE CARD,
C     WITH AN I.D. NUMBER IN COLUMNS 1-5, AND THE SCORE IN COLUMNS 6-10.  IF YOUR
C     CARDS DO NOT FIT THIS FORMAT, CHANGE THE FORMAT IN STATEMENT 5 TO SUIT YOUR
C     CARDS.  THERE SHOULD BE N SCORE CARDS.
C
      DIMENSION  F(200),  IC(200),  ICU(200),  XRANK(200),  IDN(200)
C
1     READ(5,2)   NPROB,   N,   K,   NRANK
2     FORMAT (4I5)
      READ(5,3)     (F(J),   J = 1, K)
3     FORMAT (16F5.0)
C
      J = K + 1
      F(J) = 100000.0
35    SUM = 0.0
      SSQ = 0.0
      DO 4   J = 1, K
      IC(J) = 0.0
4     ICU(J) = 0.0
C
5     FORMAT (I4,22X,F3.0)
C     THIS IS THE MODIFIABLE FORMAT FOR INPUT SCORES.
C
      DO 11  L = 1, N
      READ(5,5)    ID,   X
      SUM = SUM + X
      SSQ = SSQ + X * X
```

```
            DO 10    J = 1, K
            IF (X - F(J+1))  6, 10, 10
      6     IC(J) = IC(J) + 1
      C     INCREMENTS FREQUENCY COUNT FOR INTERVAL IN WHICH X FALLS.
            IF (NRANK)  8, 8, 7
      7     XRANK(L) = X
            IDN(L) = ID
      8     GO TO 11
     10     CONTINUE
     11     CONTINUE
            ICU(1) = IC(1)
            DO 12    J = 2, K
     12     ICU(J) = ICU(J-1) + IC(J)
      C     CREATES CUMULATIVE FREQUENCY DISTRIBUTION IN ICU(J).
      C
            EN = N
            XMEAN = SUM / EN
            VAR = (SSQ - SUM * SUM / EN) / (EN - 1.0)
            SDEV = SQRT  (VAR)
            WRITE(6,13)    NPROB, N
     13     FORMAT(36H1UNIVARIATE DISTRIBUTION. PROB. NO. I6,6H. N = I6)
            WRITE(6,14)    XMEAN, SDEV, VAR
     14     FORMAT(8H0MEAN = F13.5,10H    S.D. = F13.5,13H   VARIANCE = F13.5)
            WRITE(6,15)
     15     FORMAT(53H0FREQUENCY AND CUMULATIVE FREQUENCY DISTRIBUTION OF X)
            WRITE(6,17)
     16     FORMAT(48H0INTERVAL LOWER LIMIT    FREQUENCY   CUMULATIVE F.)
            WRITE(6,17)     (J,  F(J),  IC(J),   ICU(J),    J = 1, K)
     17     FORMAT(5X,I4,F14.6,5X,I6,5X,I6)
      C
            IF (NRANK)  1, 1, 18
     18     L = 0
            DO 21    J = 2, N
            IF (XRANK(J) - XRANK(J-1))  21, 21, 20
     20     TEMP = XRANK(J-1)
            IDT = IDN(J-1)
            XRANK(J-1) = XRANK(J)
            IDN(J-1) = IDN(J)
            XRANK(J) = TEMP
            IDN(J) = IDT
            L = 1
     21     CONTINUE
            IF (L) 26, 26, 18
      C     DATA ARE RANKED FROM LARGEST SCORE TO SMALLEST IN XRANK(J).


                       26    WRITE (6,22)
                             WRITE (7,22)
                       22    FORMAT (20H0RANKED OBSERVATIONS)
                             WRITE(6,23)
                             WRITE(7,23)
                       23    FORMAT (25H0RANK     I.D.       SCORE)
                             DO 24    J = 1, N
                             WRITE(6,25)   J,   IDN(J),   XRANK(J)
                       24    WRITE(7,25)   J,   IDN(J),   XRANK(J)
                       25    FORMAT (I4, 3X, I6, 3X, F8.3)
                             CALL EXIT
                             END
```

SUMMARY

```
C       LIST 4.3    SUBROUTINE UNIDIS
C
C       THIS SUBROUTINE COMPUTES THE MEAN AND STANDARD DEVIATION, A
C       FREQUENCY DISTRIBUTION, AND A CUMULATIVE FREQUENCY DISTRIBUTION
C       FOR A SINGLE VARIABLE, X. THE   N  SCORES ARE GROUPED INTO   K
C       CLASSES, THE LOWER LIMITS OF WHICH ARE FOUND IN  F(J). THE
C        FREQUENCIES FOR THE CLASSES ARE RETURNED IN   IC(J).
C
        SUBROUTINE UNIDIS (N, K, X, F, IC)
C
        DIMENSION    X(1000), F(100), IC(100), ICU(100)
C
        WRITE (6,2)    N
   2    FORMAT (30HOUNIVARIATE DISTRIBUTION. N =   I6)
        F(K+1) = 1000000.
        SUM = 0.0
        SSQ = 0.0
        DO 3   J = 1, K
        IC(J) = 0
   3    ICU(J) = 0
        DO 4   L = 1, N
        SUM = SUM + X(L)
        SSQ = SSQ + X(L) * X(L)
        DO 5   J = 1, K
        IF (X(L) - F(J+1))  6,  5,  5
   6    IC(J) = IC(J) + 1
        GO TO 4
   5    CONTINUE
   4    CONTINUE
        ICU(1) = IC(1)
        DO 7   J = 2, K
   7    ICU(J) = ICU(J-1) + IC(J)
        EN = N
        XMEAN = SUM / EN
        VAR = (SSQ - SUM * SUM / EN) / (EN - 1.0)
        SDEV = SQRT (VAR)
        WRITE (6,8)        XMEAN,  SDEV,   VAR
   8    FORMAT (8HOMEAN = F10.5,10H    S.D. = F10.5,13H  VARIANCE = F10.5)
        WRITE (6,9)
   9    FORMAT (53HOFREQUENCY AND CUMULATIVE FREQUENCY DISTRIBUTION OF X )
        WRITE (6,10)
  10    FORMAT(50HOINTERVAL LOWER LIMIT    FREQUENCY    CUMULATIVE F.    )
        WRITE (6,11)       (J,  F(J),  IC(J),  ICU(J),   J = 1, K)
  11    FORMAT (5X, I4,F14.6,5X,I6,5X,I6)
C
        RETURN
        END
```

CHAPTER FIVE

Computer Experiments with Random Numbers

5.1 Random number generation

Two purposes of the computer laboratory exercises in this course are to equip you with programs for computing the statistics studied and to teach you how to read and modify these programs to suit your needs. Thus, you can already compute means, variances, z scores, and T scores, and if you wanted to change the input format of one of these programs to suit the way you have your data punched, you could do so. Or, if you wanted to change the standard deviation of the T scores from 10 to 15 you would know where in the program it is set to 10 and could make the change.

Another purpose of these computer laboratory exercises is to generate large sets of certain statistics, based on many different random samples of the same population of scores, so that you may examine the resulting empirical distributions of these statistics. For example, when you draw a random sample of 100 eight-grade students from a large school system and compute the mean, that statistic is your estimate of the population mean for all eight graders from that school system. Naturally, that sample mean would probably be different from means based on other random draws of 100 students from that same population. This chapter describes a method whereby you are able to study the relationship between a large set of obtained sample means (the sampling distribution of the mean) and what we would theoretically expect that distribution to look like.

The authors believe that these sampling experiments will help you to develop conviction regarding the appropriateness and trustworthiness of the statistical procedures you are learning. These inductively acquired insights are basic to a sound understanding of statistical methodology yet do not require rigorous mathematical derivations which this book tries to avoid. In short, these computer experiments are the only "proofs" we offer to you, and it behooves you to consider their strengths and weaknesses very carefully.

To illustrate the meaning of a sampling experiment, consider the following activity which you might have engaged in if you wanted to study the distribution of sample means. Suppose you made up 2000 3 × 5 cards to represent a

RANDOM NUMBER GENERATION

population of scores. On each card you wrote a two-digit number in the range 00 to 99. There are 100 different possible whole number scores in this range, so you made up 20 cards for each of the 100 possible scores. When you were finished writing cards, you had created a population of 2000 uniformly distributed integers. You put the 2000 cards in a large box, closed it, shook and tumbled it vigorously, and then reached in blindly and drew out 10 cards. These 10 cards were a random sample of size 10 from the population. You computed and recorded the sample mean, then replaced the 10 cards in the box. Once again you shook and tumbled the box, then blindly drew a second sample of 10 cards, from which you computed and recorded a second sample mean. You repeated this sampling operation 100 times, each time replacing the previous sample of 10 cards. When you were finished, you had in your possession 100 sample means from the sampling distribution for the mean of random samples of size 10 from a uniformly distributed population of 2000 two-digit integers. What could you learn from these results?

You would be interested in the mean of your 100 sample means, so you would compute it. You would also be interested in how the sample means distributed, so you would prepare a frequency polygon to portray the distribution of means graphically. You would probably want to see how large the standard deviation of the sample means was. Now, if you knew the statistical theory that tells how the means are theoretically distributed as outcomes of such a sampling experiment, you would want to compare your empirical sampling distribution with the theoretical distribution for these means. This comparison would help you to decide whether you should trust or believe in the statistical theory.

We are going to use the computer, rather than 3 × 5 cards and a tumble box, to conduct sampling experiments for this course. The objective, as we said before, is to help you obtain personal conviction that statistical methods work the way theory claims they do. Once again we remind you that the basic purpose of statistical analysis is to obtain reasonable convictions about unknown parameters of population distributions from known properties of random samples. In actual research, the worker usually has only *one* sample to generalize from, and he has to trust statistical theory as a guide to generalizing about his data. In these sampling experiments we shall have many samples and shall be able to see how sample statistics distribute as estimates of population parameters.

Sampling experiments as described above are very tedious to stage since it is necessary to prepare a large population of scores on cards and physically draw many random samples from that population of cards. Fortunately the computer can be programmed to generate numbers so nearly random that we can accept them as if they were random numbers. This makes sampling experiments quite convenient and feasible. A set of numbers is considered "randomly sampled" if on any draw each of the possible numbers within the range has an equal chance of occurring. The population of numbers from which random numbers are generated has a uniform, or rectangular, distribution, so that every number has the same probability of occurring in the sample.

Although you are probably not familiar with the kinds of random number generators used in computers, there is one method of generating random numbers with which you are very familiar. If we were to toss one die from a pair of dice, we would generate random numbers which range from 1 to 6. The observed frequency distribution of Table 5.1 is the result of six throws of a die. The mean of that sample is 4.0. The mean of any subsequent sample would vary between 1.0 and 6.0. Theoretically, on each throw any of the six possible values of X are equally likely so the theoretical mean of die throws is 3.5. We would expect the mean of a set of means of six throws of the die to get closer and closer to this theoretical value as the number of dies is increased. By studying the distribution of sets of observed means, we can learn a great deal about means and what to say about the difference between two observed means, for example. The main point for you to see at this stage is what is meant by random number generators. Exactly how such generators will help you to understand statistics will become clearer in the remaining chapters.

Table 5.1 Using a Die as a Random Number Generator

	Observed Distribution		Theoretical Distribution	
X	f	fX	f	fX
6	1	6	1	6
5	1	5	1	5
4	2	8	1	4
3	1	3	1	3
2	1	2	1	2
1	0	0	1	1
		$\sum X = 24$		21
	$M_X = \frac{24}{6} = 4.0$		$\mu_x = \frac{21}{6} = 3.5$	

There is considerable literature on methods of generating random numbers in the computer and of testing the degree of departure from randomness of the numbers obtained from those methods.[1] The Preface of one manual of

[1] An excellent summary of the whole problem, which you would enjoy, is Chapter 9, "Random Number Generators," of Professor Green's book (Green, 1963). Another informative overview is Chapter 6, "Monte Carlo Methods," in Professor Galler's book (Galler, 1962).

RANDOM NUMBER GENERATION

such methods begins with the warning that "random numbers are harder to come by than one might suspect" (IBM, 1959).

Most experts endorse a method called the "power residue" method, which we have adopted. Our specifications for this method have been taken from the IBM Manual titled *Random Number Generation and Testing* (IBM, 1959). This manual states that the power residue method "is in many ways superior to other methods and also is entirely satisfactory if used properly." The method consists of only two elementary operations. First, an arbitrary fraction (X_0) is multiplied by (*) a selected constant (C). Second, the whole-number portion of the resulting product is discarded and the fractional part of the product is taken as the first "random" fraction (X_1). This fraction (X_1) is also used as the new multiplier of the constant to get the second "random" fraction (X_2), and so on.

A recommended constant C for a decimal operation on computers is 10011. Thus the power residue method may be specified symbollically as

$$X_j = C * X_{j-1} \text{ (modulo 1)} = 10011 * X_{j-1} \text{ (modulo 1)} \quad (5.1.1)$$

The notation "modulo 1" indicates that only the fractional remainder of a number is to be retained. X_0 is an eight-digit fraction which must be supplied to start the process.

Our function subprogram *RANDOM* (LIST 5.1), requires the user to provide his own first value. The entire series generated is controlled by the starting value selected, and a series may be reproduced exactly at any time by starting again with the same X_0. Usually this is to be avoided, but at times it may be useful. It should be noted that the most significant (leftmost) digits of the fraction are the most random, and in fact our random number routine returns only the four most significant digits of the remainder to the main program. That is, our random numbers X are in the range $.0000 \leq X_j \leq .9999$.

To take some of the mystery out of how this pseudorandom number generator works, let us trace the arithmetic by which four pseudorandom fractions are generated, using the starting fraction .53952704.

(1) $\quad X_1 = 10011 * .53952704 \text{ (modulo 1)}^2$
$\quad\quad = 5401.20519744 \text{ (modulo 1)}$
$\quad\quad = .20519744$
$\quad\quad$ and RANDOM reports $X_1 = .2051$

(2) $\quad X_2 = 10011 * .20519744 \text{ (modulo 1)}$
$\quad\quad = 2054.23157184 \text{ (modulo 1)}$
$\quad\quad = .23157184$
$\quad\quad$ and RANDOM reports $X_2 = .2315$

---

[2] The modulo 1 here means the X_1 equals the fraction portion of this product. "Modulo 1" is dropped from the equation when the whole number portion is dropped.

(3) $\quad X_3 = 10011 * .23157184$ (modulo 1)
$\quad\quad\quad = .26569024$
and RANDOM reports $X_3 = .2656$

(4) $\quad X_4 = 10011 * .26569024$ (modulo 1)
$\quad\quad\quad = .82499264$
and RANDOM reports $X_4 = .8249$

To see how a different sequence may be started, suppose the starting fraction .00100101 is read in. Then

(1) $\quad X_1 = 10011 * .00100101$ (modulo 1)
$\quad\quad\quad = 10.02111111$ (modulo 1)
$\quad\quad\quad = .02111111$
and RANDOM reports $X_1 = .0211$

(2) $\quad X_2 = 10011 * .02111111$ (modulo 1)
$\quad\quad\quad = .34332221$
and RANDOM reports $X_2 = .3433$

(3) $\quad X_3 = 10011 * .34332221$ (modulo 1)
$\quad\quad\quad = .99864431$
and RANDOM reports $X_3 = .9986$

(4) $\quad X_4 = 10011 * .99864431$ (modulo 1)
$\quad\quad\quad = .42818741$
and RANDOM reports $X_4 = .4281$

Frequently what the main program will require is a one-, two-, or three-digit integer in the range *0* through *N*, and this will be obtained by multiplying the random fraction by $N + 1$ and "fixing" the product (discarding the fractional remainder, the logical opposite of the modulo 1 operation). Or, if the desired range is *1* through *N*, this will be obtained by multiplying the random fraction by *N*, "fixing" the product, and adding *1* to this result.

To illustrate, let us convert the four random fractions we have just generated, that is,

$$.0211$$
$$.3433$$
$$.9986$$
$$.4281$$

into four random binary digits in the range $0 \leq K \leq 1$. In this case, $N = 1$ and our four random digits are

$$K_1 = 2 * .0211 = 0$$
$$K_2 = 2 * .3433 = 0$$
$$K_3 = 2 * .9986 = 1$$
$$K_4 = 2 * .4281 = 0$$

RANDOM NUMBER GENERATION

(Note that in integer multiplication only the whole-number part of the product is retained.) Or, let us convert the four random fractions to four random subscripts in the range $1 \leq K \leq 10$.

$$K_1 = 10 * .0211 + 1 = 1$$
$$K_2 = 10 * .3433 + 1 = 4$$
$$K_3 = 10 * .9986 + 1 = 10$$
$$K_4 = 10 * .4281 + 1 = 5$$

We shall frequently use this trick for creating random subscripts in the range $1 \leq K \leq N$ to randomly designate objects to be drawn from a full set of N objects.

The listing for the function subprogram *RANDOM (J)* is presented as FORTRAN LIST 5.1.[3] J is called an argument of the subprogram and is set to 1 to allow the use of the supplied starting value. If *J* is set to 1 the first random value will be read in from a data card, in the first statement calling the subprogram in the main program. Thus, to initialize the random number generator, the first call on it in the main program will be of the form

Y = RANDOM (1)

to cause the function subprogram to read a starting value from a data card. This card must have an eight-digit number in its first eight columns. Every call for the function after the initializing first call in a program employs a zero argument, which results in the use of the last random fraction generated as the multiplicand for the new random fraction generation. An example of a typical use of the function subprogram is the following loop, which has the purpose of storing in the dimensioned variable *IX(J)* 100 random integers *in the range* 0 *through* 9:

```
    Y = RANDOM (1)
    DO 15  J = 1, 100
15  IX(J) = 10. * RANDOM (0)
```

---

[3] The authors are indebted to Mr. Bary Wingersky, Project TALENT's Computer Scientist, for the FORTRAN function subprogram RANDOM (J) which implements the power residue method for the programming in this text. The modulo 1 multiplication required was not easy to accomplish in FORTRAN, and Mr. Wingersky's solution is a very ingenious bit of numerical analysis that capitalizes on the properties of integer arithmetic. It is also possible to program this method in double precision FORTRAN, but that is not available on all FORTRAN compilers, whereas this integer arithmetic is.

In statement *15*, multiplying the random fraction output of subprogram *RANDOM* by 10 moves the decimal point one place to the right. The remaining portion of the decimal fraction is eliminated by converting the floating point number to an integer, also in statement number *15*.

We should work through the actual steps performed by *RANDOM (1)* to see how the function subprogram arrives at $X_1 = .2051$ when the starting fraction read is .53952704, so that the initial values for K_1, K_2, K_3, and K_4 are 53, 95, 27, and 04. We have the following operations.[4]

M0 = 11 * K4	or	11 · 04 = 44
M2 = 11 * K3	or	11 · 27 = 297
M4 = 11 * K2 + K4	or	11 · 95 + 04 = 1049
M6 = 11 * K1 + K3	or	11 · 53 + 27 = 610
J = M0/100	or	44/100 = 0

Note no remainders in integer division.

K4 = M0 − 100 * J	or	44 − (100 · 0) = 44
M2 = M2 + J	or	297 + 0 = 297
J = M2/100	or	297/100 = 2
K3 = M2 − 100 * J	or	297 − (100 · 2) = 97
M4 = M4 + J	or	1049 + = 2 1051
J = M4/100	or	1051/100 = 10
K2 = M4 − 100 · J	or	1051 − (100 · 10) = 51
M6 = M6 + J	or	610 + 10 = 620
J = M6/100	or	620/100 = 6
K1 = M6 − 100 * J	or	620 − (100 · 6) = 20
X1 = K1	or	20.0 = 20
X2 = K2	or	51.0 = 51
RANDOM = X1 * 1.E-2 + X2 * 1.E-4	or	.20 + .0051 or .2051

which is returned as the first random fraction to the new program. Note that new values of *K*1, *K*2, *K*3, and *K*4 are left in the computer to set up the next

---

[4] In these equations, both * and · imply multiplication.

RANDOM NUMBER GENERATION

call on *RANDOM* (0). Upon the next call, operations start at statement 3 as follows:

M0 = 11 · 44 = 484
M2 = 11 · 97 = 1067
M4 = 11 · 51 + 44 = 605
M6 = 11 · 20 + 97 = 317
J   = 484/100 = 4
K4 = 484 − (100 · 4) = 84
M2 = 1067 + 4 = 1071
J   = 1071/100 = 10
K3 = 1071 − (100 · 10) = 71
M4 = 605 + 10 = 615
J   = 615/100 = 6
K2 = 615 − (100 · 6) = 15
M6 = 317 + 6 = 323
J   = 323/100 = 3
K1 = 323 − (100 · 3) = 23
X1 = 23.0
X2 = 15.0
RANDOM = .23 + .0015 = .2315

which is returned as the second random fraction, and new values of $K1$, $K2$, $K3$, and $K4$ are left in place.

Random numbers generated by our function *RANDOM* are supposed to sample a uniform distribution with the range .0000 through .9999. The uniformity of the obtained distribution should increase with increasing N, where N is the sample size. Study the program in FORTRAN LIST 5.2 to convince yourself that it should generate and report the grouped-data distribution for a sample for which N is 100 and a different sample for which N is 500. Punch and run the program, using a starting random fraction of your choice.

Then plot frequency polygons for both samples on the same sheet of graph paper, using a blue pencil for the sample of size 100, and a red pencil for the sample of size 500. Can you see a trend? Is it the one we have led you to expect? Figure 5.1 represents one set of running results obtained for this experiment. Pay some attention to the means and variances for the samples. Can you figure out what the population mean is? The expected variance is given in Section 5.2. Do not worry now about the Goodness of Fit results reported by LIST 5.2. In Chapter 9 you will learn how these results enable you to make a statistical test of the hypothesis of a uniformly distributed parent population as the source of the random fractions.

You have a right to be concerned that we have offered no proof of the

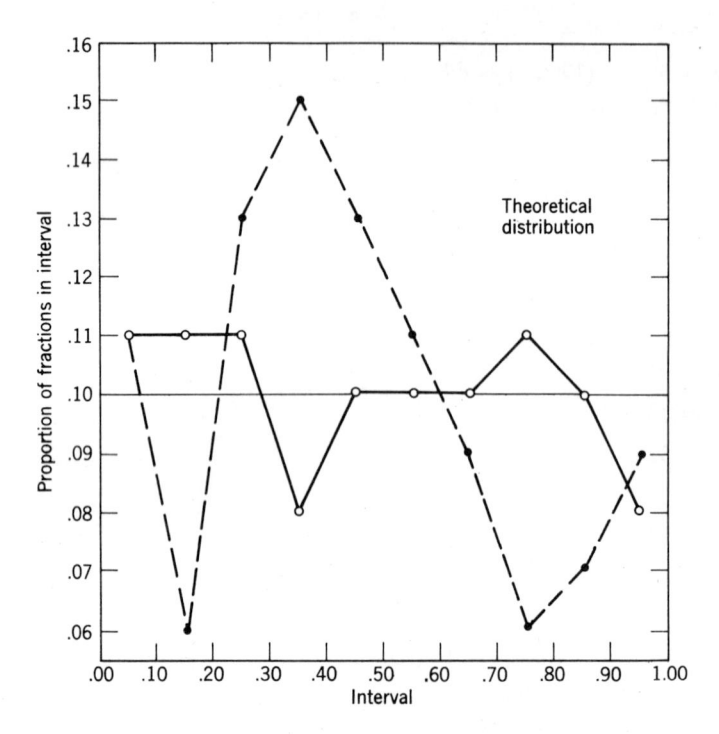

Fig. 5.1 Polygons for samples of 100 random fractions (broken line) and 500 random fractions (solid line).

approximate randomness of the fractions generated by *FUNCTION RANDOM (J)*. Even if the power residual method is heavily endorsed in the literature, there is no reason for you to assume that Wingersky's subroutine is a correct rendition of the method. You have seen that the numbers generated are reasonably uniform in distribution by inspection and have been promised that Chapter 9 will provide a statistical test of the uniformity hypothesis, but uniformly distributed numbers are not necessarily random. The fact is that it is very difficult to prove the randomness of a series of numbers. In the computerized laboratory exercises which are to come, sets of numbers generated by *RANDOM (J)* will be used to generate quite a few different empirical distributions which will be shown to fit closely to several different theoretical distributions. In every case the assumption of randomness of the output of *RANDOM (J)* will be one of the assumptions under test in the goodness of fit test. After the last of these experiments is done in Chapter 14, you will be asked to review and summarize the accumulated evidence for the randomness of the numbers generated by the random generator. By then the evidence will be pretty conclusive.

5.2 A Monte Carlo study of the variance estimates

It is customary to describe empirical studies of statistics based upon random numbers as Monte Carlo studies. This Monte Carlo study has two purposes. It seeks to demonstrate both the bias in the sampling distribution of the sample variance, $s^2 = \sum x^2/N$ and the correction provided by the unbiased estimate of the variance, $\hat{s}^2 = \sum x^2/(N-1)$. It also generates a sampling distribution of the mean, which will be analyzed in a later chapter. The method is to compute both biased and unbiased variances for 200 random samples of size 10 drawn from a uniformly distributed population of fractions with the range .0000 through .9999. The variance of this population is known to be .0833.[5] The average of the 200 computed sample variances should be less than this, while the average of the 200 computed unbiased estimates should be closer to it. The program in List 5.3 reports the sample mean, the sample variance, and the unbiased estimate of the population variance for each of 200 samples.

Summary statistics over the 200 samples which are reported are the average sample variance, the average unbiased estimate of the population variance, the average of the sample means, and the variance of the sample means. The first two are the immediate object of the experiment, and the last two become very important in Chapter 8. If each student in the class runs this experiment with a different starting random fraction, it will be possible to pool sets of 200 sample statistics to show how the average unbiased estimate of the population variance converges on the actual population parameter as the total number of estimates underlying the average increases. The starting value of the random fraction should be kept with each set of outcomes.

Table 5.2 contains the summary statistics and the frequency distribution for a set of 200 sample variances, and Table 5.3 for 200 unbiased estimates. The tables show that more than 67 percent of the 200 sample variances fell

---

[5] The interested student should consult E. Parzen (1960), p. 220, for the formula on which this expectation is based. There he will also find derived the expectation of the mean of a uniform distribution.

Table 5.2 Frequency Distribution for 200 Sample Variances of 10 Random Fractions ($M_{s^2} = .073$, $S_{s^2} = .022$)

Class	Lower Limit	Frequency	Cumulative Frequency
1	.000	4	4
2	.033	14	18
3	.043	21	39
4	.053	28	67
5	.063	35	102
6	.073	33	135
7	.083	31	166
8	.093	16	182
9	.103	10	192
10	.113	5	197
11	.123	3	200
12	.133	0	200

Table 5.3 Frequency Distribution for 200 Unbiased Estimates of Variance of 10 Random Fractions ($M_{s^2} = .081$, $S_{s^2} = .024$)

Class	Lower Limit	Frequency	Cumulative Frequency
1	.000	2	2
2	.033	7	9
3	.043	16	25
4	.053	22	47
5	.063	25	72
6	.073	34	106
7	.083	31	137
8	.093	29	166
9	.103	14	180
10	.113	9	189
11	.123	7	196
12	.133	4	200

below the population variance, while slightly more than half of the 200 unbiased estimates fell below the population variance. Compare also the average variances of the two methods, .073 and .081, with the known population variance of random fractions, .0833. These results nicely confirm the claim that the sample variance *underestimates* the population variance, and that the unbiased estimate corrects the bias in the sampling distribution of the variance. Do your own results of this *MONTE CARLO STUDY OF THE SAMPLE VARIANCE* allow the same interpretation?

5.3 Summary

Monte Carlo computer experiments employ random numbers as data to test or demonstrate statistical sampling theories.

The power residue method employs the simple formula

$$X_j = C * X_{j-1} \text{ (modulo 1)}$$

to compute a series of random fractions. In this formula, C is a suitable constant, and "modulo 1" means that only the fractional part of the product is retained as the new random fraction. The fractions generated are from a uniformly distributed population, which means that hypothetically every possible fraction has the same probability of occurrence.

FORTRAN LIST 5.1 contains a function subprogram, *RANDOM (J)*, for the generation of random fractions by the power residual method.

FORTRAN LIST 5.2 represents an experiment testing the uniformity of the fractions produced by *RANDOM (J)*, that is, the frequency distribution is rectangular.

FORTRAN LIST 5.3 represents an experiment testing the theory of the bias in the sampling distribution of the variance, and the value of the correction afforded by the unbiased estimate of the population variance.

EXERCISES AND QUESTIONS

1. How would you modify List 5.2 so that it produced random numbers in the range 00 to 99 rather than decimal fractions?

2. What is the theoretical variance for the distribution of Table 5.1, using samples of six throws of a single die? Can you design a small Monte Carlo experiment using a single die that would be a replication of the experiment in Section 5.2 ?

3. What is a common device that is used to generate random binary numbers (0 or 1)? What is the expected mean of that distribution?

4. Be sure to do the two computer exercises suggested in Sections 5.1 and 5.2.

5. This chapter speaks of "the expected variance" and "the expectation of the mean." How do you suppose this concept of *expected value* relates to the concept of a parameter?

6. From your experiences with the material of this chapter, try to formulate clearly the distinction between a *sampling distribution* and a *theoretical distribution*. This distinction figures very heavily in the work ahead and will gradually become clearer.

7. If you want to read an unusual book, consult *A Million Random Digits with 100,000 Normal Deviates* by the RAND Corporation (Glencoe, Ill.: Free Press, 1956). The willingness of scientists to pay $15.00 for these two random number tables is evidence of their widespread usefulness. RAND Corporation has distributed many copies of these tables on punch cards for computer applications. The volume opens with an informative discussion of ways of generating and testing pseudorandom numbers. One member of the class might report on the ingenious and extensive tests for randomness to which these tables were subjected.

8. What did our Monte Carlo study and comparison of variance estimates demonstrate?

```
C       LIST 5.1    FUNCTION RANDOM (K)
C
C       RANDOM NUMBER GENERATOR,  CREATED BY MR. BARY WINGERSKY.
C
        FUNCTION RANDOM (K)
C
        IF (K)   1,  2,  1
1       READ (5,3)      K1, K2, K3, K4
3       FORMAT (4I2)
        WRITE (6,4)     K1, K2, K3, K4
4       FORMAT (31H0STARTING NUMBER FOR RANDOM IS    4I3)
C
2       M1 = 11 * K4
        M2 = 11 * K3
        M3 = 11 * K2 + K4
        M4 = 11 * K1 + K3
        J = M1 / 100
        K4 = M1 - 100 * J
        M2 = M2 + J
        J = M2 / 100
        K3 = M2 - 100 * J
        M3 = M3 + J
        J = M3 / 100
        K2 = M3 - 100 * J
        M4 = M4 + J
        J = M4 / 100
        K1 = M4 - 100 * J
        X1 = K1
        X2 = K2
C
        RANDOM = X1 * 1.E-2 + X2 * 1.E-4
        RETURN
        END
```

```
C       LIST 5.2    UNIFORMITY OF RANDOM FRACTIONS EXPERIMENT
C
C       THIS PROGRAM REQUIRES SUBROUTINES RANDOM, UNIDIS, AND GOOFIT.
C       THE SINGLE DATA CARD SUPPLIES AN 8-DIGIT STARTING NUMBER FOR
C          RANDOM IN COLUMNS 1-8.
C
        DIMENSION    X(1000),  F(100),  IC(100),  IE(100)
C
        WRITE (6,2)
2       FORMAT (42H1UNIFORMITY OF RANDOM FRACTIONS EXPERIMENT    )
        CALL RANDOM (1)
        WRITE (6,3)
3       FORMAT (20H0SAMPLE 1,    N = 100     )
        N = 100
        DO 4    J = 1, N
4       X(J) = RANDOM (0)
        T = .10
        IU = 10
        F(1) = .00
        IE(1) = 10
        K = 10
        DO 5    J = 2, K
        F(J) = F(J-1) + T
5       IE(J) = IU
        CALL UNIDIS (N, K, X, F, IC)
        CALL GOOFIT (K, F, IE, IC)
C
        WRITE (6,6)
6       FORMAT (20H0SAMPLE 2,    N = 500     )
        N = 500
        DO 7    J = 1, N
7       X(J) = RANDOM (0)
        IU = 50
        DO 8    J = 1, K
8       IE(J) = IU
        CALL UNIDIS (N, K, X, F, IC)
        CALL GOOFIT (K, F, IE, IC)
        CALL EXIT
        END
```

```
C      LIST 5.3    MONTE CARLO STUDY OF THE SAMPLE VARIANCE
C
C      THIS PROGRAM GENERATES 200 SAMPLES OF 10 PSEUDO-RANDOM FRACTIONS
C      EACH, COMPUTES FOR EACH SAMPLE THE MEAN, SAMPLE VARIANCE, AND
C      UNBIASED ESTIMATE OF THE POPULATION VARIANCE.
C      REQUIRED SUBROUTINES ARE RANDOM, UNIDIS, AND GOOFIT.
C      THE SINGLE DATA CARD CONTAINS AN 8-DIGIT STARTING NUMBER FOR
C          RANDOM IN COLUMNS 1-8.
C
       DIMENSION   XM(1000), V(1000), UEV(1000),  F(100), IC(100), IE(100)
C
       WRITE (6,2)
    2  FORMAT (25H1MONTE CARLO ON VARIANCE     )
       N = 200
       K = 10
       ENDF = 9.
       EN = 10.
       X = RANDOM (1)
       DO 4    J = 1, N
       SUM = 0.0
       SSQ = 0.0
       DO 3    L = 1, K
       X = RANDOM (0)
       SUM = SUM + X
    3  SSQ = SSQ + X * X
       XM(J) = SUM / EN
       XSDS = SSQ - SUM * SUM / EN
       V(J) = XSDS / EN
    4  UEV(J) = XSDS / ENDF
C
       WRITE (6,5)
    5  FORMAT (35H0DISTRIBUTION AND FIT OF 200 MEANS    )
       K = 8
       F(1) = .00
       F(2) = .35
       T = .05
       DO 6    J = 3, K
    6  F(J) = F(J-1) + T
       IE(1) = 10
       IE(8) = IE(1)
       IE(2) = 18
       IE(7) = IE(2)
       IE(3) = 30
       IE(6) = IE(3)
       IE(4) = 42
       IE(5) = IE(4)


       CALL UNIDIS (N, K, XM, F, IC)
       CALL GOOFIT (K, F, IE, IC)
C
       WRITE (6,7)
    7  FORMAT (51H0DISTRIBUTION OF 200 SAMPLE VARIANCES                )
       K = 12
       F(1) = .000
       F(2) = .033
       T = .010
       DO 8    J = 3, K
    8  F(J) = F(J-1) + T
       CALL UNIDIS (N, K, V, F, IC)
C
       WRITE (6,9)
    9  FORMAT (53H0DISTRIBUTION OF 200 UNBIASED ESTIMATES OF VARIANCE   )
       CALL UNIDIS (N, K, UEV, F, IC)
       CALL EXIT
       END
```

CHAPTER SIX

Theoretical Distributions

6.1 An experiment

In order to examine the way in which theoretical distributions are used in research, it is useful to consider a series of hypothetical experiments in extrasensory perception (ESP). Two students are selected to serve as subjects in the experiment. A deck of playing cards is shuffled and spread out, face down, on a table separating the two subjects. One of the students serves as the "transmitter." That is, he picks up and looks at a card and thinks of its color, red or black. The other student is the "receiver," and attempts to identify the color of the card held by the student by "reading his mind."

If a card is randomly selected from the deck and it is a club, the transmitter thinks "black." The receiver then says either "black" or "red," whichever he feels is correct. His saying "black" would constitute a hit. If this receiver correctly identified three out of four cards, would you be ready to believe in ESP?

The first question that might be asked is what are the different possible outcomes of the experiment if the receiver was only guessing? Table 6.1 summarizes these possible outcomes in terms of hits and misses. The sixteen different outcomes include four ways of obtaining three hits. This means that if the receiver was simply guessing, he would have three out of four hits about one-fourth of the time (4/16 = 1/4) in repetitions of this experiment. That is, repeating the experiment many times we would expect to find that the subject could do this well 25 percent of the time even if ESP were not operating. In fact, he could do this well *or better* about 31 percent of the time (25 percent for three hits plus 6 percent for four hits) just by guessing.

If random behavior is a reasonable alternative explanation for the outcome of this experiment, an investigator who is himself convinced that ESP can and does operate in such a situation will have to provide more convincing evidence if he wishes to convince others of the existence of this phenomenon. He can be encouraged by the fact that at least the trend is in his favor, since there were more hits than misses.

One way to improve the previous experiment would be to use eight cards

Table 6.1 Possible Outcomes of ESP Experiment Using Four Cards

Hits (H) and Misses (M) Record	Number Correct	Frequency of Occurrence	Proportion of Total
Trial 1 2 3 4			
H H H H	4	1	1/16 = .06
H H H M			
H H M H	3	4	4/16 = .25
H M H H			
M H H H			
H H M M			
H M M H			
M M H H	2	6	6/16 = .38
H M H M			
M H M H			
M H H M			
H M M M			
M H M M	1	4	4/16 = .25
M M H M			
M M M H			
M M M M	0	1	1/16 = .06
	Totals	16	1.00

instead of four. It is necessary to work out a new table of possible outcomes using eight cards. One way of doing this is to write out all of the possible combinations of hits and misses as was done in Table 6.1. The student is not encouraged to try this, since there are 256 such possible outcomes. Table 6.2 summarizes the possible ways that the different number of hits could occur.

Table 6.2 Possible Outcomes of ESP Experiment Using Eight Cards

Number Correct	Frequency of Occurrence	Proportion of Total	Cumulative Proportion
0	1	.004	.999
1	8	.031	.995
2	28	.109	.964
3	56	.219	.855
4	70	.273	.636
5	56	.219	.363
6	28	.109	.144
7	8	.031	.035
8	1	.004	.004
Totals	256	1.000	

AN EXPERIMENT

In Table 6.2 you can see that there is only one way out of 256 for the receiver correctly to identify all eight cards. This means that only about four times in 1000 could this be done by simply guessing. If the color of all eight cards could be correctly identified, we would certainly tend to feel that the receiver was not guessing. There are eight ways out of 256 of having seven correct (missing either the 1st, or 2nd, or 3rd . . . or 8th card). Therefore, only about three times in 100 could this be done by chance alone. This still makes guessing an unlikely explanation, but not quite as unlikely as getting all eight correct.

Guessing six correct could happen about 11 percent of the time, and having six *or more* correct could happen 14 percent of the time. This latter percentage is obtained from the column of cumulative proportions. There is no point at which you say, "now we believe it, now we don't." Certainly the higher the number of hits, the more likely it is that guessing is not a reasonable explanation. But an experimental outcome of seven or eight hits will not convince everyone that some form of ESP is operating. Although guessing may be tentatively ruled out, there are alternative explanations for this outcome.

The skeptic may call for a revised experimental setup where visual cues between transmitter and receiver are eliminated. He may request that another pair of students be selected at random and the experiment repeated. If talking had been allowed in the previous experiment, the skeptic may request that all talking be eliminated so that verbal cues would not be possible. In other words, the general design of the experiment is reexamined so that other possible alternative explanations, such as collusion, could be eliminated.

Now the experiment is repeated as redesigned and the receiver correctly identifies all eight cards. This, the investigator might feel, will finally convince everyone. But still some skeptics persist. "What is your problem?" the investigator asks. The skeptics say, "We also want a more convincing explanation of this degree of success." The skeptics are looking for some better theory, for a set of principles, perhaps, with which they can see *how* ESP can operate. All of the statistical evidence in the world will not convince those people who feel that the researcher has an inadequate explanation of the phenomenon. In fact, they have a point, because all the researcher has established so far is that it is unlikely that chance alone accounts for the outcomes of his experiment. By saying that the results are due to ESP, he hasn't said very much about the alternatives to the chance explanation.

The main purpose of this little story about ESP experiments is to show you the general way in which theoretical distributions are used in research. You now know that theoretical distributions are useful to help the investigator see how the outcome of his experiment compares with all possible outcomes governed by a random model. If the outcome of his experiment was not unlikely, then chance factors are as good an explanation of his outcome as some specific causal theory. Also, if the theoretical distribution indicates that his outcome was unlikely (that is, probably not due to chance) then his explanation (such as ESP) has one less competing explanation, but there may be other people who prefer other causal explanations. Some may even actively seek other explanations simply because they do not like the one offered by the experimenter, even though they too are willing to reject the chance explanation based on the experimental evidence.

6.2 The binomial distribution

The reader has been left in the dark regarding the source of the entries in Table 6.2. How do we know that on the basis of chance alone the experiment should yield eight hits about four times out of each 1000 replications of the experiment, so that the probability of such a sequence occurring on any one run of the experiment is .004? The answer is in the characteristics of the experiment: (1) there are only two possible outcomes of each trial, (2) the probability of each outcome is known in advance and is constant over trials, and (3) the obtained outcome of any trial is independent of the obtained outcomes of all other trials. An experiment with those characteristics is called a *Bernoulli-type experiment*, the outcomes of which follow the *binomial distribution*. Let p stand for the known probability of a "hit" on any trial. Then $q = 1 - p$ is the probability of a "miss" on any trial. Note that an experiment is defined as a series of trials of fixed length. Let n stand for the number of trials constituting an experiment. For example, n was equal to 8 in the Table 6.2 experiment.

The terms obtained when expanding $(p + q)^n$ yields the binomial distribution, which is a probability distribution for N trials with two outcomes possible at each trial. For example, if $N = 2$, then the expansion yields three terms:

$$(p + q)^2 = p^2 + 2pq + q^2$$

In an ESP experiment using only two cards, $p = .5$ is the probability of a hit for either draw and $q = .5$ for a miss. Substituting .5 for p and q in the expansion yields:

$$p^2 + 2pq + q^2 = .25 + .50 + .25$$

The three terms of the expansion represent the probabilities of obtaining 2, 1, and 0 hits, respectively, in a two-card experiment.

With four cards, the binomial expands to five terms:

$$(p + q)^4 = p^4 + 4p^3q + 6p^2q^2 + 4pq^3 + q^4$$

THE BINOMIAL DISTRIBUTION

Again substituting 0.5 for p and q, we obtain the probability distribution for the five possible outcomes, 4, 3, 2, 1, and 0 hits, respectively.

$$(p + q)^4 = .06 + .25 + .38 + .25 + .06$$

Notice that these are the same as the probabilities obtained in Table 6.1. Notice also that the coefficients for the five terms are 1, 4, 6, 4, and 1. These coefficients of the binomial expansion were the basis for determining the probabilities of Table 6.1. Whenever $p = q = .5$ the probabilities can be determined directly from the coefficients for the different outcomes because,

Table 6.3 Pascal's Triangle (Coefficients of the Binomial Expansion)

N										Sum of the Row $\binom{n}{2}$
0					1					1
1				1		1				2
2				1	2	1				4
3			1	3		3	1			8
4			1	4	6	4	1			16
5		1	5	10		10	5	1		32
6		1	6	15	20	15	6	1		64
7	1	7	21	35		35	21	7	1	128
8	1	8	28	56	70	56	28	8	1	256

as in the $n = 4$ example, $p^4 = p^3q = p^2q^2 = pq^3 = q^4$. To express this more generally, for r hits in an n card experiment, the p's and q's of each term are given by p^rq^{n-r}, and when $p = q$, then $p^rq^{n-r} = p^n$ for all values of r, leaving only the coefficients to vary from term to term.

These coefficients are quite important in mathematics. You may have encountered them before as Pascal's triangle (Table 6.3). Successive rows correspond to successive values of n, and in each row the entries from left to right are the coefficients for expanding the binomial $(p + q)^n$. The entries in the triangle also give the number of combinations of n things taken r at a time. A usual notation for combinations is $\binom{n}{r}$. This is read, n things r at a time. If $n = 7$ and $r = 5$, then from the triangle we see there are twenty-one different combinations of five things selected from among seven things (look at row 7 and count across 0, 1, 2, 3, 4, 5). In terms of cards, there are twenty-one ways of getting five hits in seven draws (one way is to miss on the first two tosses and hit on the remaining five tosses, and there are twenty other ways).

A rule for determining the coefficients $\binom{n}{r}$ of the binomial expansion can be seen in the triangle. When $r = 0$, $\binom{n}{r} = \binom{n}{0} = 1$. The next coefficient $\binom{n}{1}$ is obtained by multiplying this first coefficient by $(n - 0)/1$. To get the coefficient for which $r = k + 1$ we multiply the coefficient for which $r = k$ by $(n - k)/(k + 1)$. Later in this section a computer program incorporating these tricks is provided, by means of which the table of probabilities can be generated for any given n, assuming $p = .5$.

Just as $p^r q^{n-r}$ serves as a general expression for determining the exponents of p and q for each term in the binomial expansion, $\binom{n}{r}$ is a general expression for each coefficient (or each entry in Pascal's triangle). The expression is evaluated in terms of factorials. You may recall that 3! is read "3 factorial," and it means $3 \cdot 2 \cdot 1$. More generally, $n!$ stands for $(n \cdot (n - 1) \cdot (n - 2) \cdots 3 \cdot 2 \cdot 1)$. You should also remember that zero factorial (0!) is equal to 1, *not* zero. The general expression for a coefficient is evaluated in terms of factorials as follows:

$$\binom{n}{r} = \frac{n!}{r!(n-r)!}$$

Thus the term for two hits (r) in four trials (N) has the coefficient:

$$\binom{n}{r} = \frac{4!}{2!(4-2)!} = \frac{4 \cdot 3 \cdot 2 \cdot 1}{(2 \cdot 1)(2 \cdot 1)} = 6$$

From the factorial expression you can see that $\binom{n}{r} = \binom{n}{n-r}$. This is important because it leads to a symmetry in the distribution of experimental outcomes in those experiments for which $p = q = .5$. The experiment of Section 6.1 is such an experiment, if we assume that chance alone governs the occurrence of a hit or a miss on any trial. It is obvious that there is a symmetry in Table 6.2. In fact, we can now work out the entries in Table 6.2, putting our two general formulas together, $\binom{n}{r} p^r q^{n-r}$. For any sequence of eight hits or any sequence of zero hits in an experiment of eight trials for which $p = .5$, the probability is:

$$\binom{n}{r} p^r q^{N-r} = \binom{8}{8}(.5)^8(.5)^0 = \frac{8!}{8!\,0!}(.5)^8(1) = (1)(.0039)(1) = .0039$$

and

$$\binom{8}{0}(.5)^0(.5)^8 = .0039$$

THE BINOMIAL DISTRIBUTION

Similarly, we may figure the probability for any sequence of seven hits or of one hit in an experiment of eight trials:

$$\binom{8}{7}(.5)^7(.5)^1 = \binom{8}{1}(.5)^1(.5)^7 = (8.000)(.5)^8 = .0312$$

You may want to figure out the other entries of Table 6.2. It helps to remember that

$$p^r q^{n-r} = p^n = (.5)^n \quad \text{for any } r$$

We see, then, that the distribution of outcomes of Bernoulli-type experiments follows a formal rule:

$$p'(X = r) = \binom{n}{r} p^r q^{n-r}, \quad 0 \leq X \leq N \quad (6.2.1)$$

This is a *probability function* rule for a family of two-parameter binomial distributions, for which p and n are the parameters. The probability p' computed by the rule is the probability that the observed number of hits will equal r. Statisticians are very interested in the corresponding *cumulative probability function* $P(X \leq r)$, which can be had by summing $p'(X = r)$ for values of X from 0 to r. The last column of Table 6.2 gives the inverse cumulative function for $N = 8$. From it we learn, for example, that the probability of six or more hits out of eight trials is .144.

The theoretical mean of a binomial distribution is given by

$$\mu_b = np \quad (6.2.2)$$

and the variance by

$$\sigma_b^2 = npq \quad (6.2.3)$$

Applying these expressions to the theoretical distribution for the eight-trial experiment of Section 6.1 results in

$$\mu_b = (8)(.5) = 4$$

and

$$\sigma_b^2 = (8)(.5)(.5) = 2$$

Is it reasonable to assert that the average outcome over many eight-trial experiments of this sort should be four hits, if indeed the outcomes are governed by chance alone? Inspection of Table 6.2 shows that four hits is the most likely outcome of an experiment, with a probability of .273, and that the distribution of probabilities is symmetric about this largest probability.

Suppose we ran an experiment similar to that of Section 6.1, but allowed a blindfolded subject to name the colors of all 52 cards in a carefully shuffled deck. The outcome is that the subject calls the correct color for 42 of the cards. What is the probability of 42 or more hits in a sequence of 52 trials, assuming

$p = .5$ for each trial? Are you willing to reject the hypothesis that this observed outcome occurred by chance alone? Why, or why not? The computer program given as List 6.1 can be run with $n = 52$ to generate the information you need to answer these questions. As it stands, the program assumes that $p = q = .5$. You might like to try to generalize the program so that it can cope with a variable p. Remember that the coefficients of the terms of the binomial expansion $(p + q)^n$ will be symmetric, but the actual terms, representing the probabilities of different outcomes, will not be symmetric, since $p^r q^{n-r} \neq p^{n-r} q^r$ if $p \neq q$. Our special interest in the family of binomial distributions for which $p = .5$ will be justified somewhat in Section 7.1 when we will demonstrate the close relationship of this family of symmetric curves for discrete variables (X is always an integer in the binomial) to a family of symmetric curves for continuous variables, called *normal* curves.

6.3 The role of theoretical distributions in statistics

The ESP experiment described in the first section of this chapter was composed of a number of trials with two possible outcomes for each trial. It was shown that when the number of trials is small, the analyst can easily list all possible outcomes of the experiment and compute the probability of each possible outcome of the experiment. Of course, he must know in advance the probability of each possible outcome of a trial. Since a possible outcome of the whole experiment is a combination of possible outcomes of a trial, the rule for combining independent probabilities is available to the analyst. Note that he must guarantee the independence of the trials in the experiment. If he cannot guarantee the independence of the trials, because he knows that the outcome of a trial is influenced by the outcomes of the previous trials, all is not lost, for rules of probability exist to cover such combinations of dependent events, but the problem is much more complicated, and we do not want to go into it here. Our point is that it is quite easy to arrive at the probabilities of all possible outcomes of a simple experiment, *if* the probabilities of the possible outcomes of a simple trial are known, and *if* the trials are independent. In the example, the number of possible outcomes of any trial was two (a success or a failure), and the probability of each was assumed to be .5 by the skeptical analyst. This assumption that the card-identifying behavior under scrutiny is random in its nature is called a *null hypothesis*. That is, the null hypothesis asserts that the observed results were a sample from a population of randomly distributed outcomes so that no other explanation (such as ESP) is needed. The believer in ESP expects an improbable outcome for the experiment under this null hypothesis, so that the assertion of randomness will be discredited. Note that when only four trials were planned, all possible outcomes of the experiment and their probabilities were easily listed. When eight trials were projected, the binomial expansion was needed to facilitate the computation of probabilities.

The statistician is enchanted with the binomial distribution because it provides a theoretical distribution for the outcomes of experiments in which there are two possible outcomes for each trial. To apply the binomial distribution to his experiment, he needs only to specify p and N. Once specified, the procedure described in Section 6.2 allows him to compute the probabilities of all possible outcomes. Usually, of course, he is concerned with the sum of the probability of the observed outcome and probabilities of all outcomes beyond the observed one, because that is the probability of all outcomes as rare as the observed one (on its side of the mode). He reasons that if the probability of the observed outcome and rarer outcomes in the same direction is very small, it is unreasonable to stick to the assumption of the null hypothesis, and he rejects it. This rejection opens the door for the assertion of some other hypothesis about the value of p. The null hypothesis, which is the tested hypothesis, asserts that the discrepancies between what theory (in this case about the value of p) says should have happened and what in fact did happen are due to the operation of *chance* or *random factors alone*. The theoretical distribution from which the statistician obtains the probability of the observed outcome of the experiment on the null hypothesis is always a part of the null hypothesis, as are all its assumptions. For example, when the binomial distribution is used, the null hypothesis assumes that there are only two possible outcomes of each trial, that the trials are independent, and that the probability of a success on a trial is given by p.

The binomial provides a useful theoretical distribution for a certain kind of experiment, but there are many other theoretical distributions of use for other kinds of experiments. For example, there is the multinomial distribution for use when the number of possible outcomes of a trial is three or some other integer larger than two. It will not be presented in this text because another distribution, the normal distribution, provides a good substitute for it. In fact, it will be shown that the normal distribution is even a good substitute for the binomial when N, the number of trials, is fairly large. The facts of the normal distribution are the major burden of the next chapter, but before considering them it may be well to discuss the role of theoretical distributions in statistics in general terms. First, we offer two definitions.

> A theoretical distribution is a hypothesis about the distribution of a random variable.

> A random variable is a rule or procedure for assigning numbers to events in such a way that for every assigned number, X, there is a probability, $P(X)$, which represents the likelihood of occurrence of a score equal to or less than X in magnitude.

The scientist is interested in variables, or measurements, per se. He makes measurements and observes their empirical distributions. He is data oriented. The statistician, on the other hand, is an applied mathematician. He is interested in finding in his mathematics appropriate theoretical distributions to match the scientist's empirical distributions. These matches are always approximate, and sometimes the approximation is very rough indeed. But even rough approximations of theory to data may be extremely useful, because theoretical distributions have known mathematical properties which may be capitalized on in analysis. We say that the statistician is interested in random variables to underline his need to specify a cumulative probability function for every variable he deals with. That is, he specifies a rule that pairs with every possible value of the variable X a corresponding value of $P(X)$. Describing a scientific variable as a random variable is really performing an act of faith. It states that the variable is believed to have a cumulative probability function. This function is never known with certainty. The statistician must exercise good judgment in determining an appropriate function from what he knows of the conditions under which the measurements are collected. He must then test the goodness of his judgment by statistical methods. He hypothesizes a theoretical distribution for the random variable. This act of hypothesizing a distribution is an elemental behavior of a statistician. Anyone doing statistics must perform this rite. This elementary textbook is going to equip you to select from among four theoretical distributions which provide appropriate cumulative probability functions for the possible outcomes of many experiments. The four distributions are the *binomial*, the *normal*, the *chi-square*, and Fisher's *t*. In advanced courses you can study other theoretical distributions which enable the statistician to cope with a variety of experimental conditions beyond the ken of this book.

The alert student will be puzzled by the emphasis in this discussion on the cumulative probability function, $P(X)$. Reviewing the previous sections on the binomial distribution, he will recall that it was possible to pair with every possible outcome of an experiment, X, a probability for that unique outcome, $p(X)$. He wonders why the random variable is not adequately defined as a rule for pairing X and $p(X)$. The answer lies in the distinction between *discrete* and *continuous* variables (of Section 2.1). Indeed, the following definition is acceptable.

> A discrete random variable is a rule or procedure for assigning numbers to events in a finite set in such a way that for every assigned number, X, there is a probability, $p(X)$, which represents the likelihood of occurrence of that score.

The trouble is that in the general case of a continuous random variable there is an infinity of possible scores, and since the sum of all $p(X)$ must equal

THE ROLE OF THEORETICAL DISTRIBUTIONS IN STATISTICS

unity, the value of any particular $p(X)$ must be zero. Although this consideration makes the probability function $p(X)$ generally useless, and forces us to the cumulative probability function, $P(X)$, it is quite instructive to pause and reflect on the nature of $p(X)$ where it is very useful, in the case of the discrete variable.

In general, every random variable implies a density function, of the form $y = g(X)$, where y is an ordinate (height) of the frequency polygon over a particular value of X on the baseline. In the case of a discrete variable, the frequency given by $y = g(X)$ can be converted to a probability, $p(X)$, by dividing $g(X)$ by the total number of observations distributed, N. Thus

$$p(X) = \frac{g(X)}{N} \qquad (6.3.1)$$

for a discrete variable. This rule says that the probability of an event or score, X, is the ratio of the number of times that event or score occurs to the total number of all events or scores distributed. Note that there are two sources of, or meanings of, $g(X)$. It may be an actual frequency count or ordinate from an empirical frequency polygon. Or, it may be computed from a theoretical rule of the form

$$y = g(X) \qquad (6.3.2)$$

which gives a hypothesized density function for X. For example, if we assume a uniform or rectangular density function for X, the rule would be

$$y = \frac{N}{k} \qquad (6.3.3)$$

where k is the number of categories or levels of the discrete variable. Then

$$p(X) = \frac{y}{N} = \frac{1}{k} \qquad (6.3.4)$$

Suppose we assume a uniform distribution of 1000 observations of X, where there are five values X may assume, namely, $X = 1, 2, 3, 4,$ or 5. What is the probability that a given X, selected at random out of the 1000 available scores, has the value 3? Since $N = 1000$ and $k = 5$,

$$y = g(3) = \frac{N}{k} = 200$$

and

$$p(3) = \frac{y}{N} = \frac{200}{1000} = .20$$

or alternately

$$p(3) = \frac{1}{k} = \frac{1}{5} = .20$$

Since the distribution is uniform, $p(1)$, $p(2)$, $p(4)$, and $p(5)$ also equal .20. Also

$$\sum_{j=1}^{k} p(X = j) = 1 \qquad (6.3.5)$$

Looking back over these remarks on theoretical distributions in statistics reveals that a second elemental behavior of statisticians has been alluded to, and may be specifically emphasized here. This is the act of inferring a probability distribution for a discrete variable from an observed frequency distribution, as one way of moving from data to theory. The statistician is not so naive as to believe that more than a crude approximation to a probability distribution can be obtained from an observed distribution of N points. What he does believe is that if the N points are randomly drawn from a population of all possible observations on the variable, the goodness of the approximation to the probability distribution improves as N increases. A famous statement of this faith is Bernoulli's theorem, which is given symbolically as

$$p\left[\left|\frac{g(X)}{N} - p(X)\right| \geq e\right] \to 0, \quad \text{as } N \to \infty \qquad (6.3.6)$$

In words, this states that the probability p, of the event in brackets approaches zero as the sample size, N, approaches infinity. The event in brackets is a discrepancy between $g(X)/N$ and $p(X)$ greater than or equal to some arbitrary e.

Perhaps you have noticed that both elemental behaviors of statisticians, the hypothesizing of a theoretical distribution and the approximating of a probability distribution from an empirical one, have involved the concept of randomness. The present ambiguousness of this term will be reduced later.

6.4 Summary

An experiment with only two possible outcomes is called a *Bernoulli-type* experiment. In a *Bernoulli-type* experiment (1) there are only two possible outcomes of each trial, (2) the probability of each outcome is known in advance and is constant over trials, and (3) the obtained outcome of any trial is independent of the obtained outcomes of all other trials. When such an experiment is repeated N times, the frequency of outcomes of each type can be accumulated and converted to proportions of the total number of outcomes, N.

If it is possible to assign a probability, p, to outcomes of the first kind on a theoretical basis (so that the probability of outcomes of the second kind is automatically $q = 1 - p$), the *binomial distribution* provides a model for a sequence of N replications of the experiment, from which the cumulative

SUMMARY

probability of the proportion of outcomes of the first kind observed and all more extreme proportions that might have been observed can be computed. This cumulative probability can be the basis for accepting or rejecting the *null hypothesis* that the observed proportion of outcomes of the first kind deviates from the expected proportion, p, only as a result of chance.

The null hypothesis that $p = .5$ is of great interest in Bernoulli-type experiments because it asserts that the outcome of any one experiment is a random affair. FORTRAN LIST 6.1 can produce the probabilities, cumulative probabilities, and inverse cumulative probabilities of all possible outcomes of a sequence of length N, for any N, assuming $p = .5$.

The binomial distribution is an example of a theoretical distribution of the discrete type. A *theoretical distribution* is a hypothesis about the distribution of a random variable. Statistical inference is always the act of testing the assumption that the outcomes of a study represent some random variable.

A *discrete random variable* is a rule for assigning numbers to events in a finite set in such a way that for every assigned number, X, there is a probability, $p(X)$, which represents the likelihood of occurrence of that score.

For both discrete and continuous distributions the following definition holds.

A *random variable* is a rule for assigning numbers to events in such a way that for every assigned number, X, there is a probability, $P(X)$, which represents the likelihood of occurrence of a score equal to or less than X in magnitude.

EXERCISES AND QUESTIONS

1. What is the probability of getting five hearts (a heart flush in poker) when dealt five cards at random from a normal deck of cards? (Remember that p here is not .5 since one-half the cards are not hearts.) Also, assume that the probabilities don't change after each card is dealt.

2. Which type of probability, $p(X)$ or $P(X)$, cannot be used with continuous distributions? Why not?

3. What is the probability that a grade 12 female will say she is "likely to go" to college full-time (response number 3) if asked the question represented by variable 8 in Appendix B?

4. What is the probability of a grade 12 boy scoring at or below his grade-sex mean on the Project TALENT information test, Part II? Do *not* assume that $P(M_X) = .50$. Under what conditions would $P(M_X) = .50$?

5. An investigator has eight students available for an experiment and he wishes to select four of them for special treatment. How many different ways are there of randomly selecting four students from among the eight?

6. Phrase a definition of the concept of inverse cumulative proportion (see Table 6.2).

7. Put Equation 6.3.5 into words. (Don't allow yourself to skip over the algebraic expressions. You can read them!)

8. Be sure to do the computer exercise outlined at the end of Section 6.2.

```
C       LIST 6.1    BINOMIAL DISTRIBUTION
C
C          PROBABILITY FUNCTION AND CUMULATIVE PROBABILITY FUNCTIONS FOR
C       BINOMIAL DISTRIBUTION WITH N OUTCOMES.  ASSUMES  P = Q = .5 .
C
        DIMENSION    PROB(999),    CUMP(999)
C
1       READ (5,2)    N
2       FORMAT (I3)
        WRITE (6,3)    N
3       FORMAT (27H1BINOMIAL DISTRIBUTION FOR I5,8H TRIALS.)
C
        EN = N
        MP = N / 2
        CONST = .5 ** EN
C       MP IS THE MIDPOINT OF THE SYMMETRICAL DISTRIBUTION.
C       CONST IS (P**R)*(Q**(N-R)) = (P**(N-R))*(Q**R) = P**N, SINCE P = .5 .
C
        COEF = 1.0
        DO 6   L = 1, MP
        R = L
        L1 = N-L
        COEF = COEF * ((EN - R + 1.0) / R)
        PROB(L) = COEF * CONST
6       PROB(L1) = PROB(L)
        CUMP(1) = CONST + PROB(1)
        PROB(N) = CONST
C       NOTE THAT CONST IS THE PROBABILITY OF ZERO OR N HITS.
C
        DO 7   J = 2, N
7       CUMP(J) = CUMP(J-1) + PROB(J)
C
        WRITE (6,8)
8       FORMAT (47H0NO. OF    PROBABILITY    CUMULATIVE       INVERSE)
        WRITE (6,9)
9       FORMAT (50H0 HITS        FUNCTION      PROBABILITY     CUM-PROB)
        NZ = 0
        WRITE (6,10)    NZ,  CONST,  CONST,  CUMP(N)
10      FORMAT(2H0 I4,  6X,F9.5,6X,F9.5,6X,F9.5)
        M = N - 1
        DO 11   J = 1, M
        J1 = N-J
11      WRITE (6,10)      J,  PROB(J),  CUMP(J),  CUMP(J1)
        WRITE (6,10)      N,  PROB(N),  CUMP(N),  CONST
C
        CALL EXIT
        END
```

CHAPTER SEVEN

Normal Curve Theory

7.1 The normal distribution

In Chapter 6 you learned that in a Bernoulli-type experiment involving N trials there are $N + 1$ possible outcomes. In the ESP card game, for example, there were five possible outcomes (0, 1, 2, 3, or 4 hits) with four cards. The statistician assigns a probability to any one of these outcomes from his knowledge of the binomial distribution of the set of outcomes. But consider for a moment the problem posed by an experiment for which there is a very, very large number of possible outcomes. Measuring the exact height of an adult male U.S. citizen with a very finely calibrated instrument would be an example of such an experiment. There are so many possible readings the investigator could get on any one male that it is reasonable to conceptualize an infinite population of available readings. Since there are an infinite number of possible readings, the probability of any one reading occurring as the result of such a measurement experiment is precisely zero. In such cases, where the experimental result is one point on a continuous scale conceived of as containing an infinite number of points, it is necessary to work with the cumulative probability of all points less than or equal to the observation. That is, we want the cumulative probability of all scores on the continuous variable, X, less than or equal to the obtained score, X_i. The cumulative probability function, $P(X_i)$, for such a continuous variable is defined as

$$P(X_i) = \text{probability}(X \leq X_i) \qquad (7.1.1)$$

It is possible to think of this cumulative probability in terms of a proportion of the area in a graph. For example, in the discrete variable case, the area covered by the bars of a histogram up to and including X_i, divided by the total area covered by the histogram, is the probability of getting a score of X_i or lower. That is $P(X_i)$. Figure 7.1 illustrates this point. The total area covered by the histogram is 10 square units. For $X_i = 2$, the cumulative probability, $P(X_i)$, is the ratio of the area covered by the first two bars to the total area. This is 2.5 square units/10 square units, so $P(X_i) = .25$. If you think of the total area as equaling 1 unit of area, the area represented by $X = 2$ or less is .25.

It is also possible to think in terms of area under a curve. The frequency polygon of Figure 7.1 illustrates that idea. Since scores between 2.00 and 3.00 are not possible for this discrete variable, the area represented by $P(X_i)$

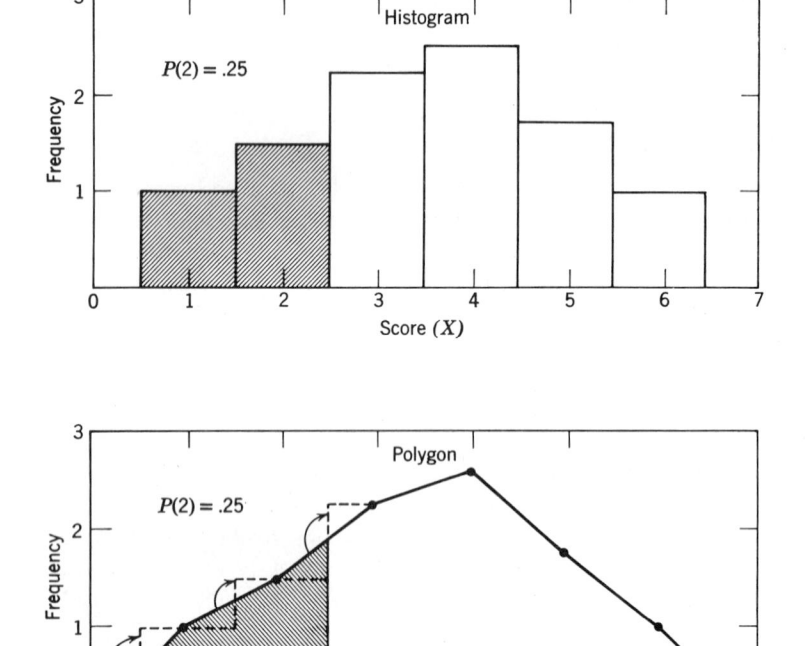

Fig. 7.1 Relationship between area and cumulative probability.

extends from 0 to 2.5 on the graph. Again the shaded area represents one-fourth of the total area under the curve. This time it was more difficult to compute $P(X_i)$ because it is easier to find areas for rectangles. However, by measuring the height of the curve at $X = 1$ and multiplying that times the base length (1 inch), and adding the height times base at $X = 2$, the area is obtained. By superimposing the histogram onto the polygon, you can see how the areas "average out" to be the same. In the graph this was done for $X = 2$ and $X = 1$.

One way of looking at the problem of determining $P(X_i)$ for a continuous

THE NORMAL DISTRIBUTION 109

variable is to view it as a problem of computing and adding up a lot of little areas. For discrete scales of only a few categories this was illustrated graphically in Figure 7.1. For continuous scales we need to go to the calculus of integrals, which provides methods for determining the area between two points on a continuous distribution. Integrals can be thought of as a procedure for adding up a lot of little areas, and that, of course, involves multiplying the height of the curve (at some point X_i) times an arbitrarily small segment of the base line around X, and adding all these little areas together to get the total area. In integral calculus, the cumulative probability $P(X_i)$ from the lowest possible point $(-\infty)$ to X_i is expressed as in Equation 7.1.2, where

$$\int_{-\infty}^{X_i}$$

stands for the sum of all the little areas between $-\infty$ and X_i; $g(X)$ stands for the height of the curve at some point along X; and dX stands for a very small length along the base line of X. The height or ordinate is written $g(X)$ because it is a function of X (that is, the height of the curve depends upon the value of X).

$$P(X_i) = \int_{-\infty}^{X_i} g(X) \cdot dX \qquad (7.1.2)$$

The concept of density is often used with $g(X)$ since the height of the curve depends upon the frequency with which observations tend to occur at various segments along X. For this reason $g(X)$ is often referred to as a density function. It is very helpful to choose a form for the frequency distribution of X for which the density function is known and is tractable in expression 7.1.2. There is one form of frequency curve for a continuous variable which is so frequently encountered in practice and so completely tractable mathematically that it has enjoyed enormous popularity in statistics and measurement work in the behavioral sciences for over half a century. This is the normal distribution, or normal curve. There is an important statistical reason for this popularity as well, which will be considered in the next chapter. Before we present all the good reasons for selecting the normal distribution as the form for the theoretical frequency distribution of a measurement variable X, we want to emphasize that this is a choice situation, and that some other frequency curve may occasionally fit the data or the requirements of theory or both better than the normal curve does. You should also know that many measurement instruments in use in psychology have been arranged to yield data which are approximately normally distributed, so that the normal-type distribution is an artifact rather than a natural phenomenon. Also, non-normal data distributions are sometimes transformed mathematically into approximately normal form. Still and all, there are a surprising number of variables, of which our speculative "height of adult male citizens of the U.S." is a good example, which do seem to be distributed approximately normally as a natural phenomenon. At least, large random samples of such variables match the theoretical normal distribution rather nicely.

The density function for the normal curve with parameters μ (the mean) and σ (the standard deviation) is given by

$$g(X) = \left(\frac{1}{\sqrt{2\pi\sigma^2}}\right) e^{-(X-\mu)^2/2\sigma^2} \qquad (7.1.3)$$

That is, given the mean and standard deviation, formula 7.1.3 allows you to compute the height, $g(X)$, of the normal curve for the score X. This expression for the height of the ordinate, $g(X)$, over a given score, X_i, in a normal distribution with parameters μ and σ is very basic to statistics. It is perhaps easier to remember in the following form:

$$g(X) = ke^{-.5\chi^2} \qquad (7.1.4)$$

where

$$k = \frac{1}{\sqrt{2\pi\sigma^2}}$$

$e = 2.72,$ the base of the natural logarithms

$$\chi = \frac{X_i - \mu}{\sigma}$$

$\pi = 3.1416$

The Greek χ is a "chi" (rhymes with pie). "Chi-square" (χ^2) is a familiar concept to statisticians and will be to you before you complete this course. Observe that χ here is equivalent to z, the standard score of Chapter 3.

It is customary to denote the assumption of a normal distribution for X with mean μ and standard deviation σ by writing $X: N(\mu, \sigma)$. The normal distribution is actually a family of curves, just as the binomial is. One particular normal curve is of such unique interest that elaborate tables of $g(X)$ have been published for it. This is when the variable has a mean of zero and standard deviation of 1 and is denoted $X: N(0, 1)$. This "normal, zero, one" curve, or "unit normal curve" (because the total area under the curve is unity) has the following *normal density function for $N(0, 1)$*;

$$g(X) = \left(\frac{1}{\sqrt{2\pi}}\right) e^{-.5\chi^2} \qquad (7.1.5)$$

in which χ^2 is simply the square of an observed standard score since $\mu = 0$ and $\sigma = 1$. A little pondering reveals that any normal distribution can be

THE NORMAL DISTRIBUTION   111

converted to this *unit normal* distribution by transforming all the scores to z scores. Because of this easy transformation, we shall assume that the density function inserted in the cumulative probability function $P(x)$ of expression 7.1.2 is the $g(X)$ of (7.1.5) in all applications. All this means operationally is that the transformation

$$z = \frac{X - M_x}{S_x} \tag{7.1.6}$$

is to be made before the cumulative probability of an observed score, X_i, is sought. The transformation itself is generally based on estimates of μ and σ from a sample of the population.

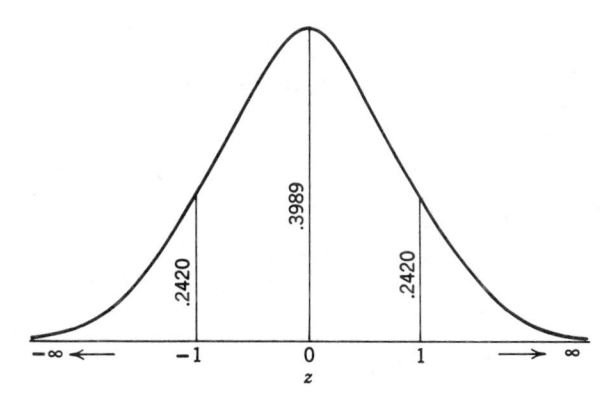

Fig. 7.2 The unit normal curve.

Figure 7.2 represents the shape of a unit normal distribution. Important features to be observed include: (1) the curve is symmetrical around the mode, that is, the left side is a mirror image of the right side; (2) the mean, median, and mode coincide; (3) the ordinate over the baseline point which is one standard deviation from the mean cuts the curve at the point of inflection; (4) the tails of the curve never actually close by meeting the baseline. There are several advantages for psychological theory in these characteristics. Suppose the measurement variable is the operational representation of a theoretical concept in the personality domain, such as verbal intelligence. The $N(0, 1)$ model suggests an infinite population of possible outcomes of testing, symmetrically distributed around a modal score of 0 in such a fashion that the density of scores falls off rapidly above 1 and below -1, but no range of scores, however rare or deviant, can be said to be totally impossible.

The values for $g(X)$ at z_i equal to -1, 0, and $+1$ are indicated on the corresponding ordinates in Figure 7.2. Use Equation 7.1.5 to reproduce at least one of these values to convince yourself that there is no mystery involved, just arithmetic. You also obtain the same result when you use Equation 7.1.3, remembering that $\mu = 0$ and $\sigma = 1$ for the $N(0, 1)$ curve.

The cumulative probability function for the unit normal curve establishes a precise relationship between any point z on the baseline of the frequency distribution and the area under the curve to the left of a perpendicular erected over that point, that is, the ordinate at that point. As we saw before, what is needed to find areas under a curve is an evaluation of the integral

$$P(z_i) = \int_{-\infty}^{z_i} g(X) \, dX \tag{7.1.6}$$

If you are unfamiliar with calculus, simply accept the statement that the integral calculus provides the means of determining the area in question. Figure 7.3 illustrates the integration problem involved. Fortunately, the need for solutions for the problem for various values of z has long since led to the preparation and publication of detailed tables of the cumulative normal probability function, for example the U.S. Bureau of Standards volume (Abramowitz and Stegun, 1964). Table 7.1 presents values of $P(z)$ for

Table 7.1 Cumulative Probabilities $P(z)$ of the $N(0, 1)$ Distribution[a]

z	$P(z)$	z	$P(z)$
−3.5	.00023	.1	.540
−3.0	.00135	.2	.579
−2.8	.00256	.3	.618
−2.6	.00466	.4	.655
−2.5	.00621	.5	.691
−2.4	.00820	.6	.726
−2.2	.0139	.8	.788
−2.0	.0228	1.0	.841
−1.8	.0359	1.2	.885
−1.6	.0548	1.4	.919
−1.5	.0668	1.5	.933
−1.4	.0808	1.6	.945
−1.2	.115	1.8	.964
−1.0	.159	2.0	.977
−.8	.212	2.2	.986
−.6	.274	2.4	.992
−.5	.309	2.5	.9938
−.4	.345	2.6	.9953
−.3	.382	2.8	.9974
−.2	.421	3.0	.9987
−.1	.460	3.5	.99977
0.0	.500	4.0	.99997

[a] Values obtained from Abramowitz and Stegun (1964), pp. 966–973.

THE NORMAL DISTRIBUTION

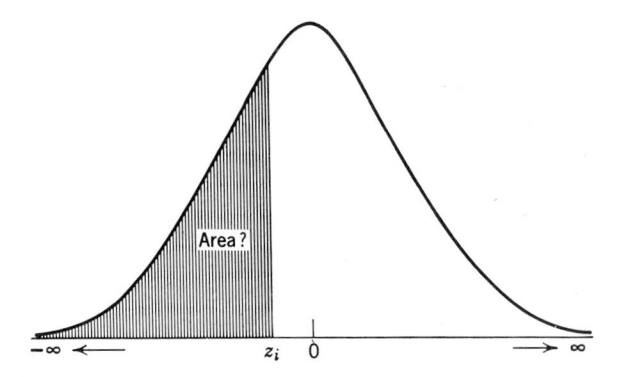

Fig. 7.3 The integration problem $P(z_i)$ depicted (total area under curve is one).

selected values of z. Examine the numbers in Table 7.1 in relation to Figures 7.2 and 7.3.

Suppose that we have selected four cases (not at random) from a complete school roster of 1000 eighth graders' z scores on a verbal intelligence test:

Pupil	z Score
Jane	-1.0
John	1.0
Jim	-1.5
Julie	2.5

Scores are standardized within that grade so that $\mu = 0$ and $\sigma = 1$. Julie has the best score in our group of four, but how good is her score in the total population of 1000 classmates? Assuming an approximately normal distribution of test scores for the class (an assumption the test manufacturer has labored to warrant for large groups of eighth graders, perhaps), Table 7.1 informs us that 99.4 percent of the 1000 scores are likely to be at or below 2.5, that is, to the left of the ordinate at $z = 2.5$. We may confidently assert that Julie is probably in the top 1 percent of her class on this test. Jim, on the other hand, has the lowest score in our panel. How poor is his position in the class? The table places him in the lowest 7 percent of the class, so that we have to expect that 93 percent of the class had better scores than Jim's, which is pretty rough on him. Both Jane and John are 1 standard deviation from the mean, but in opposite directions. What does the table lead us to say about their positions in the class? Notice that even they are pretty far out in the bottom or top of the class, and that most of the people in the class should have standard scores between minus 1 and plus 1.

Sometimes it is useful to be able to say how likely a z score of absolute

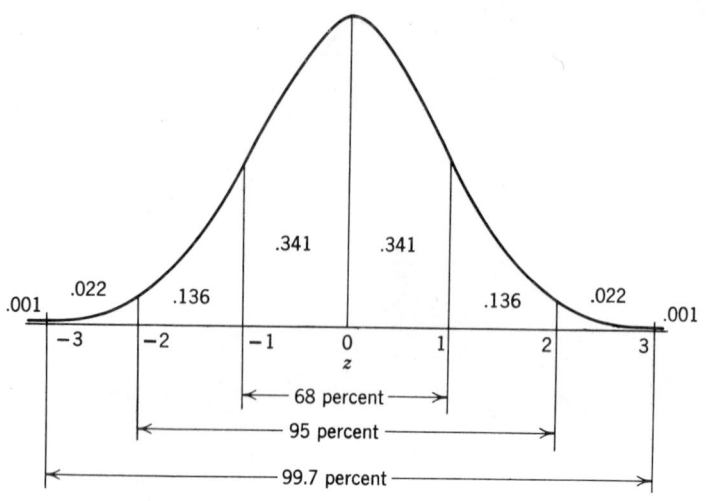

Fig. 7.4 Areas of the $N(0, 1)$ distribution.

value less than or equal to some amount is. The absolute value of Jane's score is the same as the absolute value of John's score, because in both cases $|z| = 1$. What we want to express is the probability of a z score with absolute value less than or equal to 1 in a normal distribution. Symbolically, we seek $p(|z| \leq 1)$. This probability can be derived from the entry for $z = +1.0$ in Table 7.1 by the following procedure. The table tells us that $P(+1.0) = .841$; but since $P(0.0) = .500$, we subtract .500 from .841 to get the area between 0.0 and $+1.0$ as .341; and since the curve is symmetrical about the mean of zero we reason that the area between -1.0 and 0.0 is also .341, so 2.0 times .341 gives us the required area; and we report that $p(|z| \leq 1) = .682$. We interpret this to mean that about 68 percent of the score distribution lies between standard scores of minus 1 and plus 1. Figure 7.4 represents a very informative partitioning of areas under the normal curve, and it is worthwhile to study its details.

Modifying our procedure slightly, we establish that $p(|z| > 3) = .0026$, and that $p(|z| > 4) = .00006$, so that there are approximately three chances in 1000 of a score with absolute value greater than 3.0, and only six chances in 100,000 of a score with absolute value greater than 4.0. For all practical purposes we may think of standard scores as having the range minus 4 to plus 4. Now we see that postulating a normal distribution permits a much tighter interpretation of a z score than is possible when we assume nothing about the shape of the distribution and are forced back on Tchebysheff's rule (Section 3.3). By way of contrast, Tchebysheff's rule allows 8 out of 100 scores with absolute value greater than 3.0, and 6 out of 100 scores with absolute value greater than 4.0.

THE NORMAL DISTRIBUTION

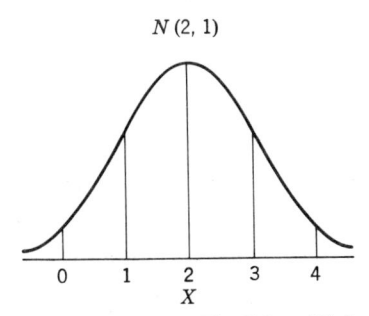

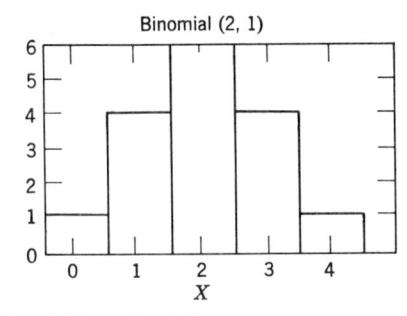

Fig. 7.5 $g(X)$ for $N(2, 1)$ and binomial $(2, 1)$.

An interesting and useful relationship between the normal distribution and the binomial distribution with $p = q = .5$ can be built via the cumulative probability functions of the two. Although the normal curve is a model for a continuous variable, and the binomial is a model for a discrete variable (see Figure 7.5), it is true that the $P(X)$ for $N(\mu, \sigma)$ is a useful approximation to the $P'(X)$ for *binomial* (μ, σ) for which $\mu = pN = .5N$ and $\sigma = \sqrt{pqN} = \sqrt{.25N}$. The approximation improves rapidly with increasing N (sample size, in this case the number of trials making up an experiment), and for small N can be improved by taking $P(X + .5)$ as the approximation to $P'(X)$. Figures 7.1 and 7.6 illustrate the reason for the *correction for continuity* provided by using $P(X + .5)$ rather than $P(X)$ in the cumulative normal function table. Note, however, that in order to use the approximation method the value $X + .5$ must be converted to a z score by substituting in

$$z = \frac{X - \mu + .5}{\sigma} = \frac{X - .5N + .5}{\sqrt{.25N}} \quad (7.1.7)$$

since our table of P is for standard scores. The data of Table 6.2 can be

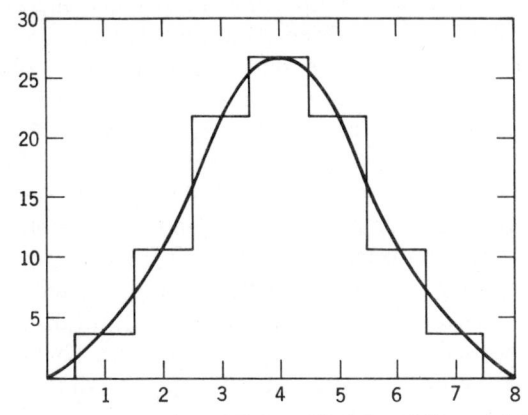

Fig. 7.6 Superimposing $g(X)$ for $N(4, 1.4)$ and binomial $(4, 1.4)$.

referred to for an example of the normal approximation to the binomial. In that example, $N = 8$, and for $X = 7$ we find that

$$z = \frac{7 - (.5)8 + .5}{\sqrt{(.25)8}} = \frac{3.5}{1.414} = 2.47$$

Table 7.1 indicates that $P(z \leq 2.5)$ is .9938. The cumulative probability for the binomial, $P'(X \leq 7)$, is shown by Table 6.2 to be .995, so even with small N we obtain a very good estimation of P' from the normal curve table. For larger N it is even better, as you will see when you do the exercises at the end of this chapter.

7.2 The central-limit theorem

The time has come to partially unveil one of the most remarkable and pervasive results in mathematical statistics. This result, called the central-limit theorem, states in part that *sums of independent random* samples of a variable tend to be normally distributed. The only restriction is that the variable sampled must have a finite mean and variance. No restriction whatever is placed on the form of the population distribution from which the samples are drawn! The seemingly outrageous assertion is made that a sum of N elements of a random sample will tend to have a normal distribution, regardless of the type of parent distribution from which the sample of size N is drawn. Notice that what the theorem postulates is the tendency of the theoretical distribution of a sum to be normal. It has already been stated in Section 6.3 that statisticians expect the empirical distribution of any statistic to approximate the theoretical distribution better as the number of empirical observations distributed increases.

That means in this case that the empirical sampling distribution of a sum would reveal the theoretical distribution better as the number of samples drawn and the number of sums computed and distributed increases, whatever the theoretic distribution is. However, the central-limit theorem itself asserts that the theoretic distribution of a sum of elements of a random sample approaches normality as N, the sample size, approaches infinity.

For example, the random number generator RANDOM(K) has been carefully engineered to produce random samples of any size from a uniform distribution of fractions in the range $.0000 \leq x \leq .9999$. The central-limit theorem proposes that the sampling distribution for a variable created as the sum of N of these random fractions would tend toward normality, that the approach to normality would increase with increasing N, and that the sum of an infinite number of random fractions from a uniform distribution would form a perfectly normal variable. In this case, the sum would be random as well as normal, since sums of random variables are known to be random also.

This theorem provides the cornerstone of statistics, as will be evident in

THE CENTRAL-LIMIT THEOREM

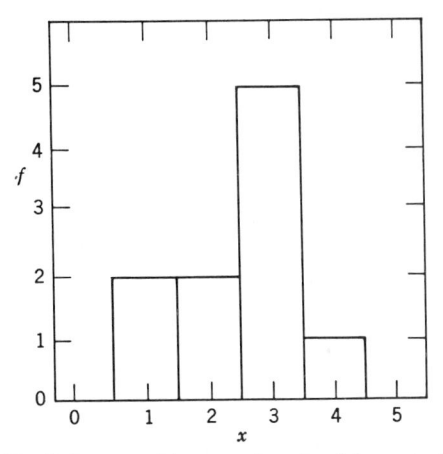

Fig. 7.7 Outcomes of ten experiments of five coin-flips each.

the next chapter, yet we cannot offer to prove it mathematically. Even the introductory mathematical statistics text we have recommended to the more mathematical student does not undertake to prove it (Mood and Graybill, 1963) because of the difficulty of the proof. We can offer a demonstration which illustrates the principle, and a computer laboratory experience which you may find quite persuasive. For our demonstration, we flipped a coin fifty times, forming ten independent samples, each containing five random elements from a uniformly distributed binary variable (assuming a true coin, and scoring a "head" as 1 and a "tail" as 0). The five trial outcomes of each sample have been summed, and the result is a nonuniform variable which exhibits a marked mode and the beginnings of some symmetry around that mode (Table 7.2 and Figure 7.7). The experiment was then repeated for another series of ten samples of five flips each, producing similar results (Table 7.3 and Figure 7.8). The reader should note that this experiment could

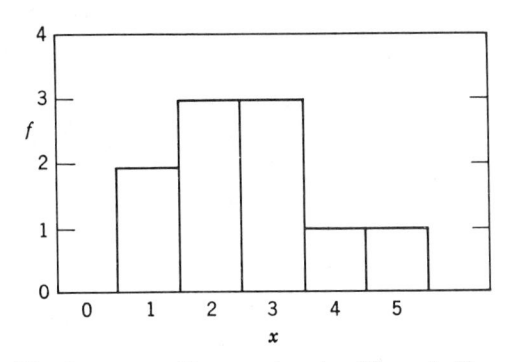

Fig. 7.8 Outcomes of ten experiments of five coin-flips each.

Table 7.2 Outcomes of Ten Experiments of Five Coin Flips Each

Experiment	Outcomes	Sum
1	0 0 0 0 1	1
2	0 1 1 1 1	4
3	0 1 0 1 1	3
4	0 1 0 1 1	3
5	1 0 1 1 0	3
6	0 1 0 0 1	2
7	0 1 1 0 1	3
8	1 0 1 1 0	3
9	0 0 1 0 0	1
10	0 0 1 1 0	2

be enlarged in two directions. It would be sensible to increase the number of samples of size five for which sums are distributed, to improve the estimation of the theoretical sampling distribution of these sums. This theoretical distribution we already know is the binomial distribution with $p = q = .5$ and $N = 5$. Figure 7.9 depicts this ideal form which our results will increasingly approximate as we increase the number of samples the sums of which we distribute empirically. This binomial distribution is as close to normality as sums of a Bernoulli-type experiment with $N = 5$ can ever go, despite the central-limit theorem and is the actual theoretical distribution of such sums. Our faith that increasing the number of sample sums distributed in our empirical approximation of this theoretical distribution will increase the goodness of the approximation is well placed and is itself an application of an implication of the central-limit theorem called the *law of large numbers*. A formal statement of this principle has been given in Section 6.3 as proposition 6.3.6, and there will be further exposition of it in the next chapter.

But consider the other way in which the experiment could be enlarged.

Table 7.3 Outcomes of Ten Experiments of Five Coin Flips Each

Experiment	Outcomes	Sum
11	0 0 1 1 0	2
12	0 1 0 1 1	3
13	1 1 1 1 1	5
14	1 0 1 1 0	3
15	0 1 1 0 0	2
16	0 1 1 1 1	4
17	0 0 1 0 0	1
18	0 0 0 1 1	2
19	0 1 1 0 1	3
20	0 0 0 0 1	1

THE CENTRAL-LIMIT THEOREM

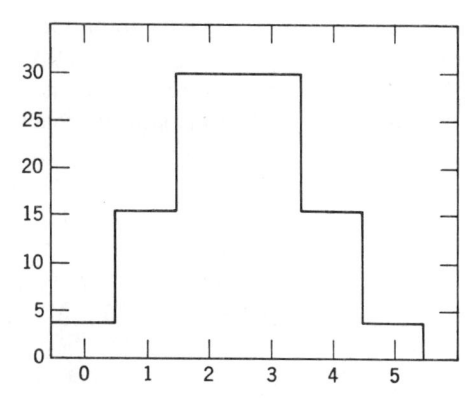

Fig. 7.9 Binomial (2.5, 1.12) for $N = 5$.

We could increase the sample size, N, which is specifically involved in the central-limit theorem as stated above. That is, we could increase the number of flips entering into each sum, for example from five to eight. The theorem explicitly guarantees that the sum of eight elements will be more normally distributed than the sum of five elements. We know that the actual distribution will now be the binomial with $N = 8$ and $p = .5$. Will this be closer to normality? Indeed it will. Compare Figure 7.9 with 7.6. A binomial based on $N = 12$ will be even more normal than one based on $N = 8$, as shown in Figure 7.10. Thus the normal approximation of the binomial with increasing

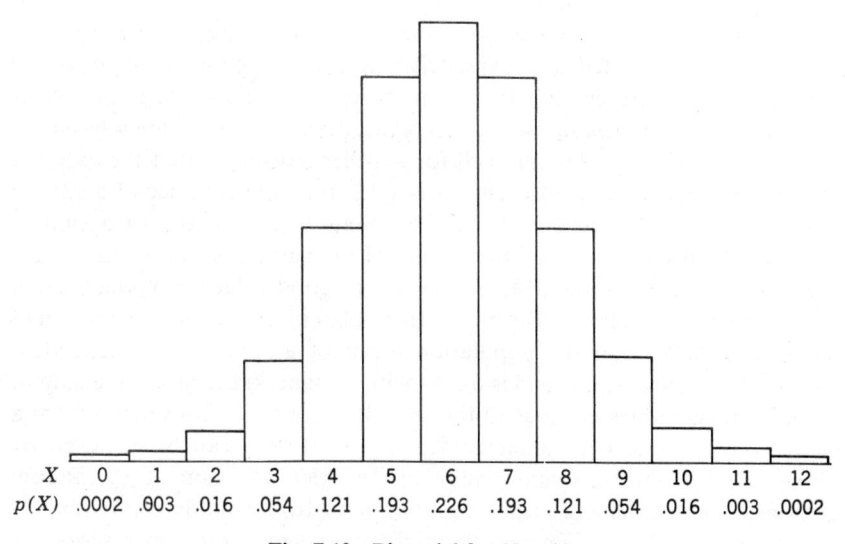

X	0	1	2	3	4	5	6	7	8	9	10	11	12
$p(X)$	.0002	.003	.016	.054	.121	.193	.226	.193	.121	.054	.016	.003	.0002

Fig. 7.10 Binomial for $N = 12$.

N is itself an application of the central-limit theorem. Now we can mention something we didn't dare to bring up before, for fear of overstretching your credulity. Suppose we have a binomial based on a biased coin. Suppose $p = .4$ and $q = .6$. Will the sum of N trials still approach normality as N increases? Yes, but the approach will be slower. The central-limit theorem allows sums from some types of populations to approach normality faster as N increases than do sums of elements from other types of parent distributions.

The central-limit theorem has been introduced in this chapter on distribution theory for two reasons. First, it helps to justify the preoccupation of applied statisticians with the normal distribution. More force for this justification will be developed in the next chapter. Second, it provides a suggestion as to how random samples from a normal distribution can be generated by the computer. Much of Chapters 8 through 11 depends on the assumption that the measurements yielded by experiments or surveys represent samples of points from normal distributions. The null hypothesis usually involves the further assumption that the samples are drawn from a particular normal distribution. It will be extremely useful to be able to generate in the computer random samples of a normal distribution, by means of which the outcomes of experiments can be simulated. The general logic of such simulation exercises will be to show that the statistical theory works in practice when the assumptions of the particular model are met by the data. One such Monte Carlo study has already been incorporated in Chapter 5. The Monte Carlo study of the correction for bias in the sample variance did not require random sampling of a normal distribution. The actual distribution randomly sampled was a uniform or rectangular one.

The central-limit theorem suggests that a *sum* of a series of random fractions will be distributed approximately normally. Experience has revealed that 12 fractions are enough to sum in most applications, as the resulting approximation to normality will be quite adequate. This is a judicious choice for the length of the series summed for another reason, in that the variance of a uniform distribution of fractions is 1/12, thus the variance of a sum of 12 such fractions is equal to 1. This is because the variance of a sum of independent variables is equal to the sum of the variances. Since unity is the ideal variance for a normal distribution, 12 is a good value for N, the number in the series to be summed. The population mean of a uniform distribution of fractions is .5000, and the population mean of a sum of 12 independent fractions is 6 (since 12 times .5 is 6). As with variances, the mean of a sum of independent variables is equal to the sum of the means. Conversion from a population mean of 6 to a more ideal mean of zero can be achieved by subtracting 6 from each sum generated. In brief, the sum of 12 random fractions, minus 6, should yield a random point from what is approximately an $N(0, 1)$ distribution. The result might be called a random z score, or *random normal deviate*. List 7.1 contains a FORTRAN function subprogram called RANDEV that reports such a deviate.

How can you test the assertion that the score reported by RANDEV(O) is acceptable as a random normal deviate? Using the List 7.2 program, you could generate 1000 supposed random normal deviates, and generate the cumulative proportions and compare with values of $P(z)$ provided in Table 7.1. Please do this work and see if the results are convincing. Of course, you do not expect to match the tabled values exactly. How can you explain slight departures of the results of this experiment from the expected cumulative distribution for a random normal deviate? In Chapter 9 you will learn a statistical test for the hypothesis that the departures of the distribution of

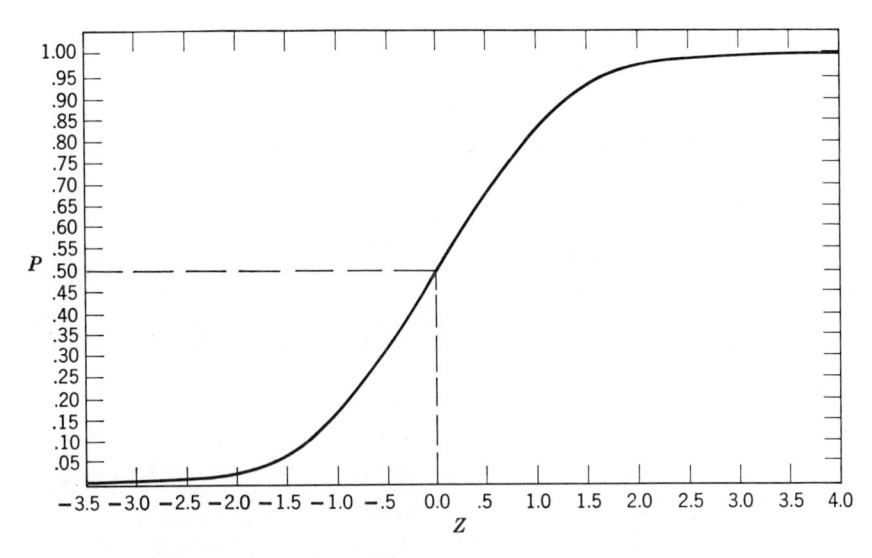

Fig. 7.11 Ogive for $N(0, 1)$ (with plotted points representing observed $P(Z)$ for Table 7.4).

your 1000 sample points from the theoretical distribution are attributable to chance alone. Save your data distribution for later application of this "goodness of fit" test.

A graphic plot of the cumulative normal probability function is called the *normal ogive*. The curve of Figure 7.11 represents the normal ogive. In Figure 7.11 we have also located points based on 1000 deviates computed by the authors using the starting random fraction .00000011. The points are based on the results reported in Table 7.4. Do this same thing for your results.

Table 7.4 Grouped-Data Frequencies and Cumulative Frequencies for 1000 Random Normal Deviates Generated by RANDOM(J), List 7.2

Interval	Lower Limit	Observed f	Theoretical f	Observed cf	Theoretical cf for $N(0, 1)$	Observed $P(Z)$
1	−4.00	0	1	0	1	.000
2	−3.00	27	22	27	23	.027
3	−2.00	120	136	147	159	.147
4	−1.00	361	341	508	500	.508
5	−.00	332	341	840	841	.840
6	1.00	134	136	974	977	.974
7	2.00	24	22	998	999	.998
8	3.00	2	1	1000	1000	1.000
9	4.00	0	0	1000	1000	1.000

7.3 Departure from normality

Psychologists usually prefer to assume that human or animal traits are distributed approximately normally, rather than to go out on a limb and assert an absolutely normal distribution. What they do is to suppose that the true distribution of a characteristic is probably close enough to normal to make it useful to apply normal curve analyses to sample data on the characteristic. There are several ways in which an observed distribution curve may be different from normal.

Reminding ourselves that the normal curve is symmetrical, it is easy to imagine distributions that are not quite symmetric. *Skewness* is the name for such a departure from normality. One type of skewness, known as positive skew, looks like this

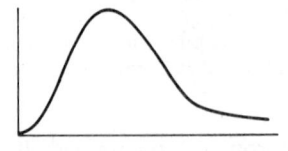

Income in the United States is positively skewed in its distribution. The other type of skewness is negative skew, which looks like this

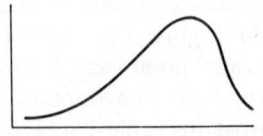

Age at death for Americans is negatively skewed.

SUMMARY

The other characteristic of a normal curve that is subject to violation is the requirement that inflection points exactly coincide with 1 standard deviation above and below the mean. Technically, a point of inflection is a place where the rate of acceleration changes sign and is located by calculus. What is at issue from a nontechnical point of view is the flatness or peakedness of the curve as compared with the normal curve with the same parameters. This aspect of the shape of a curve is termed *kurtosis*, and a curve may have too much or too little kurtosis to qualify as normal. The curve with too much kurtosis is peaked and may look like this

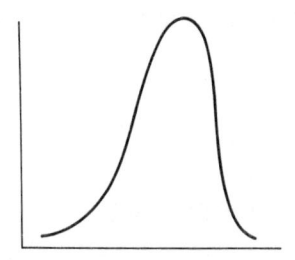

Such a curve is said to be leptokurtic. The curve which errs in the other direction, in that it has too little kurtosis and is too flat is said to be platykurtic. It may look like this

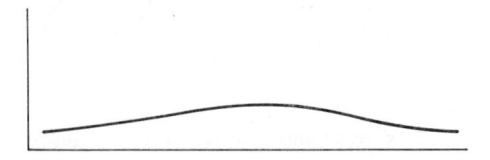

There are statistical indices and tests for skewness and kurtosis which we shall not explore. When you see mention of "an approximately normal distribution" you will know the ways in which a unimodal distribution curve may depart from the normal curve that has the same mean and standard deviation.

7.4 Summary

Behavioral scientists favor the normal curve as a distribution theory for their measurement variables because it is intrinsically attractive in its symmetry, unimodality, and open-endedness; because it is mathematically tractable; and because it is the natural distribution for many traits and for errors of measurement, as we shall see later.

The areas of the unit normal curve between the mean and 1, 2, and 3 standard deviations above or below the mean are well worth memorizing. Figure 7.4 displays them graphically.

The normal curve probabilities for standard scores give good approximations to the binomial probabilities for reasonably large N. The observed frequency of hits, X, must first be converted to standard score form:

$$z = \frac{X - \mu}{\sigma} = \frac{X - pN}{\sqrt{pqN}}$$

For smaller N, a correction for continuity is obtained by adding .5 to the deviation score.

One version of the *central-limit theorem* asserts that sums of independent random samples of a variable tend to be normally distributed. FUNCTION RANDEV(K) uses this theorem to construct random normal deviates out of sums of uniformly distributed random fractions.

EXERCISES AND QUESTIONS

1. Using List 4.1, compare some of the variables of Appendix B with the theoretical normal. Use either sex. (We suggest variables 6, 8, and 9.)

2. Work out the thirty-six possible outcomes for the sums of a pair of dice and show the theoretical distribution of the thirty-six throws (using the eleven possible sums as class intervals). Notice how this distribution of sums is closer to bell-shaped than if you had a single twelve-sided die for which all sides were equally likely.

3. For the binomial with $N = 52$ and $X = 42$, compute the cumulative probability $P(X)$ using the normal curve approximation method of Section 7.1. Compare this result with the computer-derived result for the binomial which you did in Chapter 6.

4. Be sure to do the exercise suggested at the end of Section 7.2.

5. Assuming that the distribution of Information, Part I (R-190) scores of Appendix B.2 (males) is normal, determine $P(X)$ for student with $ID = 45$ (where $X = 141$). Compare this cumulative probability with the one you obtain from the cumulative distribution of the List 4.2 program, which you probably used to obtain M_X and S_X anyway ($M_X = 156.6$).

6. Although the binomial distribution approximates the normal distribution for large N, there is still a basic theoretical difference between the two distributions. What is this major difference?

7. What proportion of a large set of normal deviates would be expected to fall between $z_1 = -1.80$ and $z_2 = +1.20$?

SUMMARY

```
C     LIST 7.1    FUNCTION RANDEV(K)
C
C     RANDOM NORMAL DEVIATE GENERATOR.  REQUIRES FUNCTION RANDOM(K).
C
C     CENTRAL-LIMIT THEOREM IS USED TO GENERATE AN APPROXIMATE RANDOM NORMAL
C     DEVIATE FROM THE SUM OF 12 PSEUDO-RANDOM FRACTIONS.
C
C     PRECEDE IN CALLING PROGRAM BY STATEMENT
C           Y = RANDOM(-1)
C     AND PROVIDE INITIAL VALUE OF 8-DIGIT RANDOM FRACTION ON A DATA CARD.
C
      FUNCTION RANDEV(K)
C
      RFSUM = 0.0
      DO 1   J = 1, 12
 1    RFSUM = RFSUM + RANDOM(0)
      RANDEV = RFSUM - 6.0
C
      RETURN
      END

C     LIST 7.2    TEST FUNCTION RANDEV(K)
C
C     THIS PROGRAM GENERATES AND DISTRIBUTES 1000 PSEUDO-RANDOM NORMAL
C     DEVIATES.
C     THE SINGLE DATA CARD CONTAINS AN 8-DIGIT STARTING NUMBER FOR
C         RANDOM IN COLUMNS 1-8.
C     REQUIRED SUBROUTINES ARE RANDOM, RANDEV, UNIDIS, AND GOOFIT.
C
      DIMENSION   Z(1000),  F(100),  IC(100),   IE(100)
C
      WRITE (6,2)
 2    FORMAT (32H11000 RANDOM NORMAL DEVIATES            )
      Y = RANDOM (1)
      N = 1000
      DO 3    J = 1, N
 3    Z(J) = RANDEV (0)
      K = 8
      F(1) = -9.00
      F(2) = -3.00
      T = 1.00
      DO 4    J = 3, K
 4    F(J) = F(J-1) + T
      IE(1) = 1
      IE(2) = 22
      IE(3) = 136
      IE(4) = 341
      IE(5) = IE(4)
      IE(6) = IE(3)
      IE(7) = IE(2)
      IE(8) = IE(1)
      CALL UNIDIS (N, K, Z, F, IC)
      CALL GOOFIT (K, F, IE, IC)
      CALL EXIT
      END
```

CHAPTER EIGHT

Standard Errors and Statistical Inference

8.1 Standard error of the mean

A good deal of the fundamental theory of statistics has been presented in the first seven chapters. The result is that you have been introduced to quite a lot *about* statistics, although the only practical applications you have been shown thus far are descriptive indices such as mean and standard deviation, and the analysis of outcomes of Bernoulli-type experiments, to which you could apply the binomial distribution theory you have studied. The time has come to learn how to apply normal theory to the outcomes of surveys or experiments for the purpose of making statistical inferences.

Statistical inference is the name of the game in which the researcher generalizes beyond the set of numbers he has in hand. That is, after computing a sample statistic he may wish to make an assertion or inference about the corresponding unknown value in the population. If a teacher administers a test to twenty students and she wants to know how the average person in her class did, she may compute the mean of that class. If all she is interested in is the description of her class no statistical inference is involved. It is only when she wishes to go beyond her class mean to make statements about populations of students that she needs to call on statistical inference.

If the principal wants to know the mean of the students in his high school and does not have the time or money to administer the test to all students, he can randomly select eighty students and test them. When he has computed the mean for the eighty students tested, he can estimate the high school mean from the sample selected and tested. In order to determine the confidence he can have in his computed statistic, he has to know the manner in which the mean would vary on repeated samples from that same population. That is, he needs to know the sampling distribution of the mean. You will now be shown how and why the normal distribution is useful in this game of inference.

A random variable was defined as a rule for assigning cumulative probabilities $P(X)$ to all numbers in the domain of X. We have a random *normal* variable when the rule is the cumulative density function produced by substituting the normal distribution Equation 7.1.3 into Equation 7.1.2. We have seen that the mean and the variance are the parameters of a normal

STANDARD ERROR OF THE MEAN

distribution, so that estimates of them are the most appropriate statistics to compute on a sample of points from a normal population. We will now see some reasons why statisticians use these two statistics whenever they deal with continuous variables, regardless of the distributions involved. We are going to show how to estimate the variance of the sampling distribution of the mean of such a sample, and how to place "confidence limits" on the sample mean. These limits specify a range of values within which the population parameter has a stated probability of existing. You will see how hypotheses about the value of a population mean can be tested through confidence limits around the sample mean, and how a hypothesis regarding the difference between two populations may be tested through confidence limits on the difference between two sample means. Although the sampled distributions may not be normal, it will emerge that the sampling distribution of means tends to normality, making normal curve theory very useful indeed. Following this chapter's presentation of the logic of inferences regarding means, the next chapter extends the same logic to questions about proportions.

Computation of the mean for each of a large number of random samples of size N from a population with mean μ produces an empirical sampling distribution of the mean. In a Chapter 5 laboratory exercise you computed a Monte Carlo study of the sampling distribution of the variance. One of the unused results of that exercise was 200 sample means, each estimating the same population μ, each from a different random sample of size 10. The value of μ for the population sampled is .500. What was the average value for your sample of 200 means? On one replication of this experiment the authors happened to have an average of 200 sample means of .5008, which isn't too bad. The results which you computed probably demonstrate as nicely as ours do that the sample mean is an unbiased statistic, in that the average of its sampling distribution approaches the population mean, μ, as the number of samples increases. If you ran the experiment several times, so that you have in hand several samples of 200 means, each based on a different initial random fraction, you should find that the pooling of these samples produces an average mean even closer to μ.

Usually the research worker has only one sample mean, and he must accept it as his best estimate of the unknown population parameter μ. If he knew the variance of the sampling distribution of means based on his sample size he could make a decision about the degree of confidence he should have in his estimate. In short, he knows that his sample mean is his best estimate of μ, but he wishes to know how good an estimate of μ is M_x for a sample of size N. Suppose we let σ_M^2 stand for the variance of M_x. Notice that we are now thinking of M_x itself as a random variable, the distribution of which has parameters μ and σ_M^2. It is possible to prove mathematically that the best estimate of σ_M^2 is given by $\hat{s}_x^2/N$, where $\hat{s}_x^2$ is the unbiased estimate of the variance of X. Thus we define "the variance error of the mean" as

$$s_M{}^2 = \frac{\hat{s}_x{}^2}{N} \qquad (8.1.1)$$

From this we get the "standard error of the mean" as

$$s_M = \sqrt{\frac{\hat{s}_x{}^2}{N}} \qquad (8.1.2)$$

Applying these concepts to the Monte Carlo study of Chapter 5, where $\sigma_x{}^2 = .0833$ for the uniformly distributed variable sampled, we find that $\sigma_M{}^2 = .00833$ for samples of size 10. Take the reported average of your 200 sample values of $\hat{s}_x{}^2$ as the best available estimate of $\sigma_x{}^2$ and divide it by 10. What is the value of $s_M{}^2$ for your experimental results? The authors obtained an average $\hat{s}_x{}^2$ of .0828 on their 200 samples and from this can estimate $s_M{}^2 = .00828$, which isn't too far from the known value of $\sigma_M{}^2 = .00833$. What did you compute as the actual variance of your 200 means? The author's 200 means had an actual variance of .0076. If you will consult the program for the Monte Carlo study of the variance, you will note that this is a sample variance, not an unbiased estimate. The corresponding unbiased estimate would be .0076 also, however. Whether 1.5200 is divided by 200 or by 199 makes little difference. The correction for bias is only of practical importance in the case of small samples.

An obvious implication of Equation 8.1.1 is that the variance error of the mean decreases with increasing sample size. In fact, as $N \to \infty$ (as N approaches infinity), $\sigma_M{}^2 \to 0$ and $M_x \to \mu$. Statisticians say that M_x is an unbiased estimator of μ, and that it is a *consistent* statistic (more observations lead to better estimation). It is the *law of large numbers* which assures us that as sample size increases the probability of a given discrepancy between M_x and μ decreases. Formally this law may be stated in this application as

$$P(|M_x - \mu| \geq k) \to 0 \quad \text{as} \quad N \to \infty \qquad (8.1.3)$$

for any constant k. Compare this statement of the law of large numbers applied to a continuous variable with the application to a discrete variable in Bernoulli's theorem, expression 6.3.6. Can you see the two applications of the same principle?

Chapter 7 capitalized on a special version of the central-limit theorem to support the claim that the sum of independent random samples of a variable tends to normality. We can now benefit from a more usual statement of the theorem which asserts that the sampling distribution of M_x based on samples of size N tends to normality as N increases. The beauty of this theorem is that it places no restrictions on the distribution of the random variable X itself.

STANDARD ERROR OF THE MEAN

Regardless of the form of the distribution of X, we are assured that means based on samples of size N will be approximately normally distributed for large enough N. Conventionally an N of about 30 or more is thought to be large enough. You have from Chapter 5 a collection of 200 means based on an N of 10, where the observations were randomly sampled from a uniformly distributed population of decimal fractions. In Chapter 9 you will learn a method for testing the hypothesis that the sampling distribution of these means is the normal distribution. Later you will have acquired from Chapter 11 an alternative hypothesis for the sampling distribution of M_x based on N

Table 8.1 Grouped-Data Frequencies and Cumulative Frequencies for 200 Means of 10 Random Fractions, from List 5.3 ($\mu_M = .5000$, $\sigma_M = .0913$)

Interval	Lower Limit (M)	Actual f	cf	Lower Limit (z)	Theoretical f	cf
1	.00	9	9	−.00	10	10
2	.35	12	21	−1.64	18	28
3	.40	32	53	−1.10	30	58
4	.45	38	91	−0.55	42	100
5	.50	35	126	0.00	42	142
6	.55	32	158	0.55	30	172
7	.60	25	183	1.10	18	190
8	.65	17	200	1.64	10	200

of 10, which can also be tested. Right now we suggest that you apply some commonsense criteria to the distribution of your 200 sample means, following the techniques of Chapter 2 for describing such a distribution.

In Table 8.1 we array the actual frequencies in each of the intervals of our 200 means against the expected frequencies for a normal distribution. The expected frequencies are obtained by first converting the lower limit for each interval of M_x to a standard score, where

$$z = \frac{M_x - \mu}{\sigma_M} = \frac{M_x - .500}{.0913}$$

Entering a normal curve table with z, the $P(z)$ for the lower limit of the second interval times 200 is the cumulative frequency (cf) for the first interval. The lower limit z for the third interval yields the $P(z)$ for the second interval, etc. Once the 10 cf's are obtained, the f's are simply the cf differences from interval to interval. In Figure 8.1 the observed f's are plotted along with the expected (theoretical) f's.

Using our class intervals and theoretical frequencies, make a similar tabulation for your data and save it for use in the laboratory exercise of Chapter 9, when it will be the basis for a statistical test of the normality of

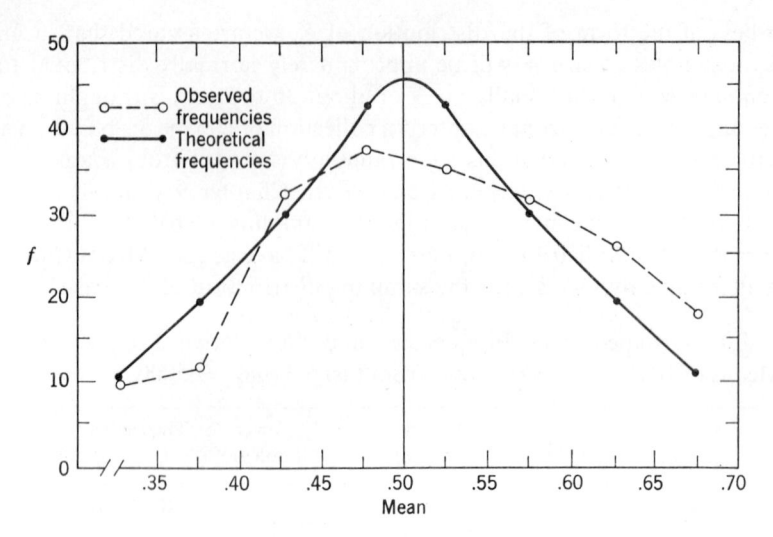

Fig. 8.1 Polygon for 200 means of 10 random fractions.

your distribution. Naturally, we do not expect the observed frequencies to match the theoretical frequencies exactly, even if it is true that M_x for samples of size 10 is normally distributed. Why not? The question we would like to be able to answer is whether the discrepancies between the observed and the expected distributions are too large to be attributed to sampling error. In other words, must we reject the null hypothesis that our 200 sample means

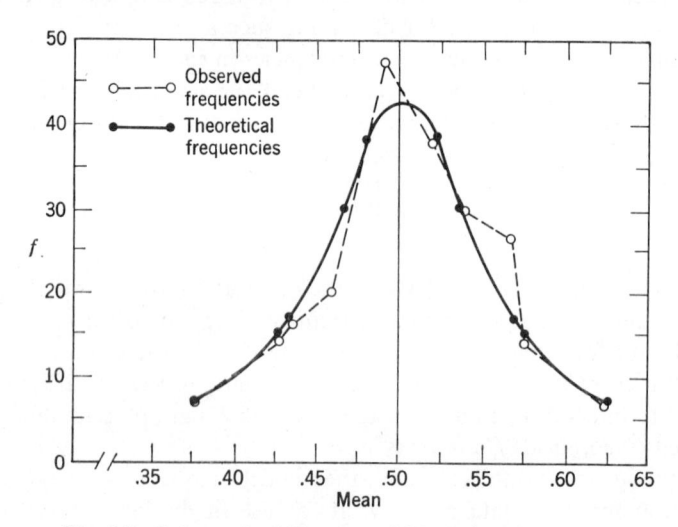

Fig. 8.2 Polygon for 200 means of 30 random fractions.

STANDARD ERROR OF THE MEAN   131

Table 8.2 Grouped-Data Frequencies and Cumulative Frequencies for 200 Means of 30 Random Fractions, from List 8.1 ($\mu_M = .5000$, $\sigma_M = .0527$)

Interval	Lower Limit (M)	Actual f	cf	Lower Limit (z)	Theoretical f	cf
1	.000	14	14	−.00	16	16
2	.425	16	30	−1.42	18	34
3	.450	20	50	−0.95	30	64
4	.475	42	92	−0.47	36	100
5	.500	35	127	0.00	36	136
6	.525	30	157	0.47	30	166
7	.550	26	183	0.95	18	184
8	.575	17	200	1.42	16	200

are a random sample from a normally distributed population of means? In Chapter 9 you will learn how to test this hypothesis.

This is a good time to test the assertion of the central-limit theorem that means based on more observations will be more normally distributed than with fewer observations. FORTRAN LIST 8.1 generates 200 means based on samples of size 30, and 200 means based on samples of size 60. Again, the observations are randomly sampled from the population of uniformly distributed fractions (that is, rectangular distributions). Compute this experiment, your frequency polygons (ours appear as Figures 8.2 and 8.3), and compare the actual frequencies in class intervals with those expected if the sample means are normally distributed as $N(\mu_M, \sigma_M)$ as we have in Tables 8.2 and 8.3. Save these results for later testing of the normal distribution hypothesis. Now, what do you think about the trustworthiness of the central-limit theorem as a result of this computing adventure?

Table 8.3 Grouped-Data Frequencies and Cumulative Frequencies for 200 Means of 60 Random Fractions, from List 8.1 ($\mu_M = .5000$, $\sigma_M = .0373$)

Interval	Lower Limit (M)	Actual f	cf	Lower Limit (z)	Theoretical f	cf
1	.000	5	5	−.00	4	4
2	.425	9	14	−2.01	14	18
3	.450	27	41	−1.34	32	50
4	.475	63	104	−0.67	50	100
5	.500	43	147	0.00	50	150
6	.525	34	181	0.67	32	182
7	.550	15	196	1.34	14	196
8	.575	4	200	2.01	4	200

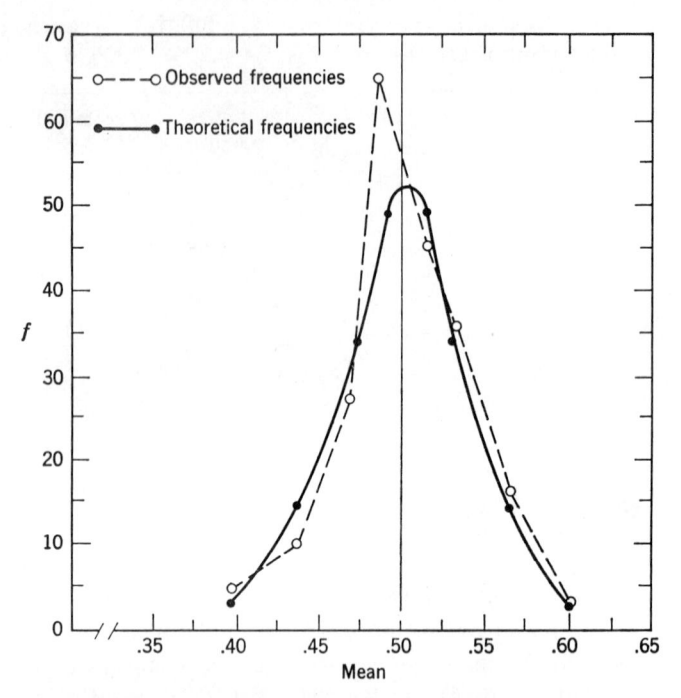

Fig. 8.3 Polygon for 200 means of 60 random fractions.

8.2 Testing a hypothesis about a mean

Knowing that the sampling distribution of the mean is approximately normal for large enough N, and that it centers on μ with variance $\sigma_M^2 = \sigma_x^2/N$, suggests that any given sample mean can be visualized relative to its position on a normal curve, as in Figure 8.4. The figure implies that the obtained sample mean (M_x) is one and a half standard errors above the

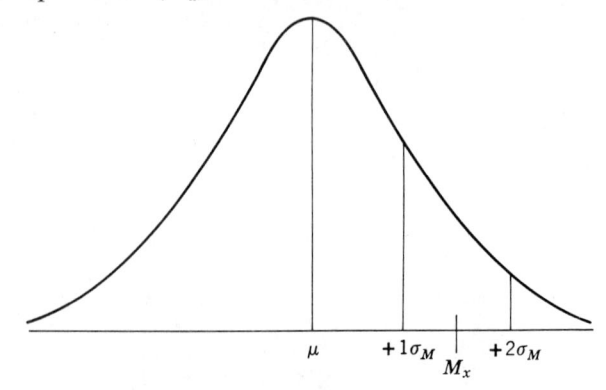

Fig. 8.4 A sample mean and its parent distribution.

population mean. At first glance, the figure seems rather useless and unconvincing, because ordinarily the parameters μ and σ_M are not available, and the whole purpose of the sample statistics M and $\hat{s}_x$ is to provide estimates of these unknowns. The figure becomes useful if we consider the situation in which there is a theoretical, or possibly empirical basis for hypothesizing a value for μ prior to collecting the data on a sample. It is then desirable to test the hypothesis that the sample arose as a random sample from a population with parameter μ, so that the deviation of M from μ can be attributed to sampling error. To get at the logic employed, consider the question, "What is the probability of the sample event M on the hypothesis of a random sample from a population with parameter μ?" A necessary assumption is that σ_M is adequately estimated by $\hat{s}_x/\sqrt{N}$. With this assumption, the standard score for M becomes

$$z_M = \frac{M - \mu}{S_M} \qquad (8.2.1)$$

The resulting z score cannot be given a probabilistic interpretation directly, since we have seen that in a continuous distribution $p(z)$ is zero regardless of the magnitude of z. It is possible to get $P(z) = p(z' \leq z)$ by looking up z in Table 7.1. This act would be unwarranted because it capitalizes on the sign of the discrepancy of M from μ, when as far as we know that sign may be due entirely to chance. We need to focus on the absolute value of the discrepancy from the hypothesized parameter value, and ask the more refined question, "What is the probability of a discrepancy as large as or larger than the observed one, regardless of sign, assuming M is based on a random sample of size N from a population with mean μ?"

This appears to be an excessively complicated question, but actually it is the simplest question for our purpose. We pool the probabilities of a value as large as ours in absolute terms and larger because (1) we would make the same decision about the hypothesis if the sign of our discrepancy happened to be reversed, and (2) we would make the same decision if the magnitude of our discrepancy (and the resulting z) happened to be greater. So, we seek the cumulative probability of all observations as extreme as or more extreme than our particular $|M - \mu|$. In Figure 8.5 the shaded areas represent the basis for the desired pooled probability. Now that we know what we want, we can compute it with the help of Table 7.1.

First, looking up $z = 1.5$ in Table 7.1 reveals that $P(z) = .933$. Subtracting .500 for the area under the lower half of the curve leaves .433 as the area between the mean of zero and 1.5, and subtracting this from .500 leaves .067 as the area to the right of, or above 1.5. The area to the left of -1.5 must be the same since the curve is symmetrical about zero, so pooling reveals that the answer to our question is $p = .134$. We could have done the same job more simply if we had observed in Table 7.1 that $P(-z) = .0668$, and doubled this to get the required $p = .1336$. The easy way, then, is to look up in Table 7.1, $P(-|z|)$, and compute

$$p(H_0) = 2P(-|z|) \tag{8.2.2}$$

The notation $p(H_0)$ stands for "the probability of the null hypothesis." In the present case, the null hypothesis asserts that the discrepancy $M - \mu$ arose by chance when M was computed on a random sample of size N from a population with parameter μ. The outcome of our computation, stated numerically as

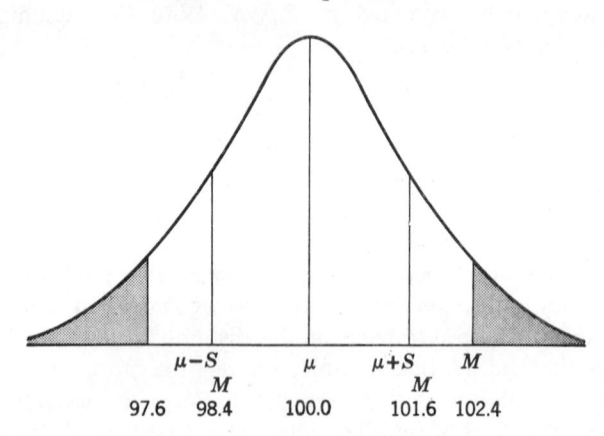

$$M = 102.4, \quad s_x = 16.0, \quad N = 100, \quad S_M = 1.6, \quad \mu = 100$$
$$z_M = 1.5, \quad p(H_0) = .13$$

Fig. 8.5 A sample mean and its estimated parent distribution, given a hypothesized $\mu = 100$, $N = 100$, $\hat{s}_x = 16$.

may be stated verbally as, "The probability of a discrepancy, without regard to its sign, of a sample mean from the hypothesized parameter of at least the observed magnitude, given a random sample of this size, is .13. That is, there are 13 chances in 100 samplings of observing a z score for the sample mean as large as or larger than this one in absolute value. Thus, the sample mean obtained does not deviate from the hypothesized parameter sufficiently to cause us to reject the null hypothesis. As far as we are concerned, the most parsimonious explanation of the observed discrepancy is to attribute it to "chance."

If we refuse to run the risk of an erroneous rejection of the null hypothesis when the odds are 13 out of 100, what risks will we accept? It has become conventional in behavioral science to accept risks of 5 or fewer chances out of 100. Ordinarily, researchers studying human behaviors would reject the null hypothesis if $p(H_0) \leq .05$. This preconceived readiness to reject the hypothesis when its probability is .05 or less is termed *the .05 alpha level*, and is reported symbolically in research journals as $\alpha = .05$. In other words, the alpha level for a research project is the largest computed $p(H_0)$ for which the researcher intends to reject the null hypothesis. It is a statement of the risk he is willing to take. Somtimes the .05 alpha level is described as *the .95 confidence level*, because when $p(H_0) \leq .05$, there are at least 95 chances out of 100 that the null hypothesis is erroneous, and we may speak of the existence of 95 percent confidence in the rejection of the hypothesis. In recent years there has been a trend among researchers to reevaluate the .05 level as perhaps too risky for maturing behavioral sciences, and the .01 alpha level (or, the .99 confidence level) is increasingly popular.

The thoughtful student will realize that whether a given deviation of a statistic from its hypothesized parameter will lead to $p(H_0) \leq \alpha$ is partly a function of sample size, N, since standard errors decrease as N increases, and smaller standard errors yield larger z scores for a fixed discrepancy. A sufficiently large N will lead to a rejection of the null hypothesis for almost any fixed discrepancy of the sample statistic from the hypothesized parameter. Some critics of statistical inference as an epistemology believe that this fact makes statistically analyzed research simply a numbers game, in which the researchers with the largest samples of subjects will enjoy the most "rejections." The authors have expressed their concern regarding research publications which report only "rejections" without consideration of magnitudes of estimated discrepancies (called "effects" in experiments) in Chapter 15. We fervently acknowledge that large samples can lead to rejections of hypotheses when the discrepancies, though real and not due to chance, are trivially small. This is a danger in interpretation of research findings which the wise researcher guards against by careful consideration of the theoretical and practical importance of his estimated real discrepancies, or effects. On the other hand, there is no substitute for statistical inference as a safeguard against interpreting as real discrepancies those which are attributable to chance, and critics who are preoccupied with the hazards of statistical inference need to give more attention to its irreplaceable value in scientific method. They also need to note that the standard error of a statistic is a function of the dispersion of observations in the population sampled, as well as of sample size. In a well-planned research, a large sample size may be the result of anticipation of large variances for observations, and may be fully justified. Certainly it is sensible that research findings based on more observations should be more reliable, other things being equal, and this is the real meaning of the relationship of sample size to standard error. An exchange of views on this matter can be found in two letters to *American Psychologist*, Bauer (1964) "versus" Cooley and Jones (1965).

The method you have just learned for testing $H_0: M - \mu = 0$ by computing

z_M and comparing $2P(-|z|) = p(H_0)$ with α is a classical example of *statistical inference*. It encapsulates a complete inductive behavior, beginning with a question about a set of observations and emerging with a generalization stated in the language of probabilities. For the given example, see if you can phrase the question and the answer yourself. It is very important that you be able to do so. Now, to test your command of this logic further, see if you can phrase a null hypothesis and a research question about a sample event on the null hypothesis for the Bernoulli-type experiment of Chapter 6, Section 6.1. Review Section 6.2 and note that it provides a method for testing H_0 in a Bernoulli-type experiment by means of the binomial distribution. Is it obvious to you that in order to get $p(H_0)$ as a two-tailed probability from the output of the FORTRAN program for the binomial distribution you will have to double the reported cumulative probability of events at least as rare as the observed event in its particular direction of deviation from the expected, or most likely, event?

But, you may say, in experiments like that of Section 6.1 there is no need to consider the odds on events in one tail of the distribution of possible outcomes, the "worse than chance" tail, since only outcomes significantly better than chance can lead to a rejection of the null hypothesis under these circumstances. If you have been this insightful you have stirred up a hornets' nest. Indeed it is possible to apply a "one-tailed test" to a null hypothesis, and in many research situations, including that of Section 6.1, it is the more appropriate test. The researcher chooses a two-tailed test of a null hypothesis when, as in our example based on Figure 8.5, he cannot anticipate from theory the direction of the discrepancy from hypothesis, and merely wishes to know whether an observed discrepancy should be attributed to chance. He suspects that the hypothesized parameter is incorrect, but he has no notion as to whether it is too large or too small. He is actually in a very weak position, but if there is strength in numbers, he may be comforted by the hordes of researchers with whom he shares this stance. Far too many behavioral scientists are satisfied with two-tailed tests of hypotheses in their researches! In time you will see that such hypothesis tests are expressions of ignorance, and only appropriate to the exploratory phases of theory building. When theory dictates the direction of the discrepancy from the hypothesized parameter, the researcher chooses a one-tailed test of the null hypothesis.

To clarify the distinction, consider Figure 8.5. It represents a situation in which a two-tailed hypothesis test is appropriate because the same decision about μ would result from an observation of $-z_M$ as results from z_M. Therefore, the significance test questions the significance of discrepancies at least as large as the observed one regardless of sign. But suppose that $M = 102.4$ is the observed mean intelligence test score for a class of 100 students, and the geographical location of the class compels the researcher to

TESTING A HYPOTHESIS ABOUT A MEAN

expect a class mean *above* the national average of 100. Now his research hypothesis is not simply the ignorant assertion that the null hypothesis is wrong. His research hypothesis is that his sample is from a population of above-average intelligence. The null hypothesis remains that M arose from a random sampling of a population with mean μ by chance, but with the new alternative hypothesis only a value of M sufficiently *above* μ can lead to a rejection of H_0. Testing now requires the use of

$$\text{One-tailed } p(H_0) = 1 - P(z), \quad \text{given } z > 0 \tag{8.2.3}$$

Applying this one-tailed test to our example, $P(z_M = 1.5) = .933$, and $1 - P = .067$. We interpret this finding as indicating seven chances out of 100 of an M at least 2.4 score points above μ on the basis of chance alone. This does not make the .05 alpha level and does not produce a rejection of the null hypothesis, but does come closer than did our two-tailed test, a point which will be explored in Chapter 9. Right now, perhaps (8.2.2) should be restated with a tag to remind us of the distinction we are laboring:

$$\text{Two-tailed } p(H_0) = 2P(-|z|) \tag{8.2.4}$$

For the eight-card experiment of Section 6.1, the null hypothesis would assert that the observed number of hits arose by chance in a random sampling of a population of outcomes with expectation, or mean, $Np_b = 8(.5) = 4$. Suppose the observed outcome is six hits. A two-tailed test would be based on the pooled probabilities of six or more hits (.144) and of two or fewer hits (.144), and would conclude that $p(H_0) = .29$, or that there are almost three chances out of ten of a discrepancy from expectation at least as large as the observed one without regard to direction on the basis of chance alone. Hence, no rejection. But, you say, the alternative hypothesis of extrasensory perception requires *more* hits than the expected four from the null hypothesis that $p_b = .5$. So, a one-tailed test looks only at the probability of six or more hits, which is .144. Again, no rejection, but again closer to the .05 alpha level than is the two-tailed probability.

Computing z_M has led us to tests of an *a priori* hypothesized μ, but at the expense of obscuring the most powerful research result, which is that *a posteriori* the most likely value of μ is M. After the findings are in, possession of a maximum likelihood estimate of μ in M, and a dispersion estimate for it in S_M, permits the construction of *confidence limits* around the mean. Since the sample mean is approximately normally distributed, it follows that about 68 percent of sample means fall within the limits $\mu - \sigma_M$ and $\mu + \sigma_M$, or that about 95 percent of sample means fall within the limits $\mu - 2\sigma_M$ and $\mu + 2\sigma_M$. (Figure 7.3 is a basis for these assertions.) We don't know μ and σ_M, but we can employ M and S_M, and argue that our best estimate of the confidence limits for μ is given by $M \pm kS_M$, where $k = 1$ for 68 percent confidence limits and $k = 2$ for 95 percent confidence limits. Thus, we argue that there are 95 chances out of 100 that μ lies within the confidence limits, or in the *confidence interval*, given by $M - 2S_M \leftrightarrow M + 2S_M$. The great value of this line of reasoning is that it dramatizes that *all* values of μ within these limits are tenable in the light of the available evidence. In our example, the 95 percent confidence limits for $M = 102.4$ and $S_M = 1.6$ indicate a range of tenable values of μ from 99.2 to 105.6. The authors conclude that the best *a posteriori* judgment regarding μ for the population sampled is the statement: "The confidence interval of 99.2 to 105.6 has a probability of .95 of including the population mean, although the best estimate of the mean value is 102.4." Such a statement nicely summarizes *all* the information about μ available from the sample data. It tells us what our best *a posteriori* (that is, after the data are analyzed) estimate of μ is, and it also tells us what the interval is within which we would expect to bracket μ 95 times out of every 100 times we create such intervals.

Another example of this type of reasoning can be drawn from the sampling study of New Hampshire high school sophomores reported in Table 3.2. The table indicates that the sample of 404 college preparatory students has a mean SCAT Verbal score of 289.2, and a standard deviation of 11.4. What are the 95 percent confidence limits for this mean estimate? The arithmetic is:

$$M \pm 2S_M \quad \text{or} \quad M \pm 2\left(\frac{S}{\sqrt{N}}\right)$$

$$289.2 \pm 2\left(\frac{11.4}{\sqrt{404}}\right)$$

$$289.2 \pm 2(.567)$$

$$288.1 \leftrightarrow 290.3$$

The interpretation is: "The probability is .95 that the interval 288.1 to 290.3 contains the mean SCAT Verbal score for New Hampshire public high school students in 1964. The best estimate of the mean is 289.2."

It may help you to think about confidence intervals if different wording is used in describing the probability involved. That is, "5 percent of the time that I establish the 95 percent confidence limits about an obtained sample mean, the population mean will *not* lie within the specified limits." Of course in the long run your assertions about the confidence limits for sample means will be correct about 95 percent of the time using $2S_M$ as the basis for estimating those limits.

8.3 Significance of difference between two means

One of the most useful statistical acts is the making of decisions regarding the significance of the difference between two sample means. In surveys it is

SIGNIFICANCE OF DIFFERENCE BETWEEN TWO MEANS 139

frequently found that subjects are naturally divided into two groups, as for example males and females, and a comparison of group means on a particular measurement is obviously attractive. The TALENT data of Appendix B can be used to illustrate this method. The male and female sample means for the mechanical reasoning test are 14 and 9, respectively. Is it likely that the sample means will be the same even if the population means happen to be exactly equal? Clearly the operation of sampling error in the selection of random samples from these populations having equal means will lead to an observed difference between sample means. When we do not know whether the population means are equal we must have a procedure for deciding whether the mean difference for samples is too large to be attributed to sampling error. In these cases we assert the null hypothesis $H_0: \mu_1 = \mu_2$ and apply statistical methods for computing $P(H_0)$. Again, the strategy is to reject H_0 if $P(H_0) \leq \alpha$, a predetermined rejection region.

In the 12th grade Project TALENT data of Appendix B, the means are based on 271 girls and 234 boys. To get $p(H_0)$ requires the standard error of a mean difference for samples of these sizes. In other words, we have to conceptualize a sampling distribution for mean differences based on random samples of sizes $N_G = 271$ and $N_B = 234$, and inquire about the standard deviation of that sampling distribution, $\sigma_{M_G - M_B}$. The null hypothesis assumes $\mu_{M_G - M_B} = 0$. For large samples, an appropriate estimate of the standard error $\sigma_{M_1 - M_2}$ is

$$S_{M_1 - M_2} = \sqrt{S_{M_1}^2 + S_{M_2}^2} \qquad (8.3.1)$$

Armed with this standard error, it is easy to compute

$$z_{M_1 - M_2} = \frac{M_1 - M_2}{S_{M_1 - M_2}} \qquad (8.3.2)$$

For large samples this statistic will be approximately normally distributed (another fruit of the central-limit theorem), and it can be referred to the table of the unit normal curve for interpretation. Once again a two-tailed test will be based on

$$\text{Two-tailed } P(H_0) = 2P(-|z|) \qquad (8.3.3)$$

In the Project TALENT example,

$$N_G = 271, \quad M_G = 9, \quad \hat{S}_G^2 = 12, \quad S_{M_G}^2 = .05$$
$$N_B = 234, \quad M_B = 14, \quad \hat{S}_B^2 = 13, \quad S_{M_B}^2 = .06$$
$$S_{M_G - M_B} = .3, \quad z_{M_G - M_B} = 16, \quad P(H_0) < .001$$

Large samples lead to small standard errors, which allow even small mean differences to emerge as statistically significant, but the mechanical reasoning difference is clearly significant.

It is legitimate for sampling surveys to be fact finding in purpose, and for their tests to be two-tailed. On the other hand, experiments are created to test hypotheses derived from theories, and almost always should pose a research hypothesis counter to the null hypothesis, so that tests can be one-tailed. The contrast of means for two independent groups, constituted by random assignment of subjects to groups, and then differentiated by exposure to different treatments, is one of the simplest classic experimental designs.

It is possible and desirable to translate findings about mean differences from independent samples into confidence intervals. The 95 percent confidence interval is constructed as

$$(M_1 - M_2) \pm 2S_{M_1-M_2}$$

Such an interval summarizes all the information about the mean difference available from the samples. Note that all the values in the interval are tenable hypothetical values for the population mean difference, although $M_1 - M_2$ is the *best* estimate of that difference. If zero falls within the interval the null hypothesis cannot be rejected at the .05 confidence level.

Another example of this reasoning about the difference between means can be drawn from the New Hampshire study reported in Table 3.2. The difference between the SCAT Verbal mean for the college preparatory sample and the corresponding mean for the nonpreparatory sample is 15. What are the 95 percent confidence limits for this difference? In Section 8.2 we computed a standard error for the college preparatory mean of .567. We now compute a standard error for the nonpreparatory mean of .587. Can you justify this number? Our additional arithmetic is:

$$M_1 - M_2 \pm 2S_{M_1-M_2}$$
$$15 \pm 2(\sqrt{.314 + .348})$$
$$15 \pm 2(.81)$$
$$13 \leftrightarrow 17$$

The interpretation is: "The probability is .95 that the interval 13 through 17 contains the difference between SCAT Verbal means for college preparatory curriculum students and noncollege preparatory students in New Hampshire public high schools in 1964."

The best estimate of the difference in means is 15.

8.4 Summary

The *central-limit theorem* assures an approximately normal distribution of sample means for large samples. The mean, μ_M, of the sampling distribution of means is the population mean for the variable, μ_X. The variance of the

SUMMARY

sampling distribution of means, σ_M^2, is σ_X^2/N. The estimate of μ_M is M_X, and the estimate of σ_M^2 is $S_M^2 = \hat{S}_X^2/N$.

List 8.1 generates means for 200 samples for which N is 30, and means for 200 samples for which N is 60. Means for 200 samples for which N is 10 are already available from running List 5.3. The frequency polygons for these three sets of sample means demonstrate the tendency toward normality with increasing N. (Refer to Figures 8.1, 8.2, and 8.3.)

The null hypothesis that a sample mean differs from a theorized population value only by chance can be tested with a *one-tailed test* by

$$P(H_0) = 1 - P(z), \quad \text{given } z > 0$$

or with a *two-tailed test* by

$$P(H_0) = 2P(-|z|)$$

where

$$z = \frac{(M - \mu)}{S_M}$$

A one-tailed test is preferable whenever the researcher has an *a priori* basis for asserting the side of μ on which he expects M to fall.

When N is about 30 or larger, *confidence limits* for μ can be constructed from sample evidence by asserting that there are 95 chances out of 100 that

$$(M - 2S_M) \leq \mu \leq (M + 2S_M)$$

All values for μ within these limits are tenable in the light of the available evidence.

When N is about 30 or larger, the null hypothesis that two population means are equal ($H_0: \mu_1 - \mu_2 = 0$) can be tested by computing

$$z_{M_1-M_2} = \frac{M_1 - M_2}{S_{M_1-M_2}}$$

where

$$S_{M_1-M_2} = \sqrt{S_{M_1}^2 + S_{M_2}^2}$$

This z score is referred to the normal curve, in either a one-tailed or a two-tailed test.

Findings on mean differences can also be stated in terms of confidence limits on $(\mu_1 - \mu_2)$ by employing

$$(M_1 - M_2) - 2S_{M_1-M_2} \leq (\mu_1 - \mu_2) \leq (M_1 - M_2) + 2S_{M_1-M_2}$$

If the confidence limits for the mean difference contain zero, the null hypothesis that the population mean difference equals zero cannot be rejected with 95 percent confidence.

EXERCISES AND QUESTIONS

1. Given the following data for a test on verbal fluency, a normally distributed variate:

	Boys	Girls
Mean	60	64
Standard deviation	12	8
N	1000	1000

   (a) About how many girls might be expected to make scores of 72 or above?
   (b) About how many girls might be expected to make scores between 60 and 72?
   (c) Who most likely scored highest, a boy or a girl?
   (d) Do you have all the information needed to test the significance of the difference between boys and girls on this variable?

2. Scores on a numerical ability test were obtained for 182 children of different socioeconomic backgrounds (SES). The following results were obtained:

	High SES	Low SES
Mean	148	128
Standard deviation	20	20
N	101	81

   What percentage of low SES students will have scores above the mean of the high SES students, if the distributions are normal? Would you be willing to attach significance to this observed difference?

3. Compute the 95 percent confidence limits for the difference between the Creativity mean for the population of grade 12 boys and for grade 12 girls using the TALENT data. Write an interpretation of the limits. From your results, what can you say about the probability of the null hypothesis that $\mu_B - \mu_G = 0$? (Most of the data analysis needed here can be done by UNIVARIATE DISTRIBUTION.)

4. Be sure to do the experiment that is outlined at the end of Section 8.1, using List 8.1.

5. What is the probability of drawing a sample with M_X equal to 51 or higher from a population with $\mu_X = 50$ and $\sigma_X = 10$ using samples of size 40?

6. If a colleague of yours criticised you for using normal curve theory in answering Question 3 without your knowledge whether or not the variable X was normally distributed, how would you answer him?

SUMMARY

```
C       LIST 8.1    MONTE CARLO ON SAMPLE MEAN
C
C       THIS PROGRAM COMPUTES AND DISTRIBUTES 200 SAMPLE MEANS BASED ON
C       30 PSEUDO-RANDOM FRACTIONS EACH, AND 200 SAMPLE MEANS BASED ON
C       60 PSEUDO-RANDOM FRACTIONS EACH.
C       THE SINGLE DATA CARD CONTAINS AN 8-DIGIT STARTING NUMBER FOR
C           RANDOM IN COLUMNS 1-8.
C       REQUIRED SUBROUTINES ARE RANDOM, UNIDIS, AND GOOFIT.
C
        DIMENSION   XM(1000),  F(100),  IC(100),  IE(100)
C
        WRITE (6,2)
 2      FORMAT (25H1MONTE CARLO ON MEANS       )
        X = RANDOM (1)
        K = 8
        N = 200
        F(1) = .000
        F(2) = .425
        T = .025
        DO 3    J = 3, K
 3      F(J) = F(J-1) + T
C
        L = 30
        WRITE (6,4)       N, L
 4      FORMAT (26H0DISTRIBUTION AND FIT FOR    I4,16H MEANS BASED ON    I3,
       C 10H FRACTIONS   )
        EL = L
        IE(1) = 16
        IE(8) = IE(1)
        IE(2) = 18
        IE(7) = IE(2)
        IE(3) = 30
        IE(6) = IE(3)
        IE(4) = 36
        IE(5) = IE(4)
        DO 6    J = 1, N
        SUM = 0.0
        DO 5    M = 1, L
        X = RANDOM (0)
 5      SUM = SUM + X
 6      XM(J) = SUM / EL
        CALL UNIDIS (N, K, XM, F, IC)
        CALL GOOFIT (K, F, IE, IC)

        L = 60
        WRITE (6,4)       N, L
        EL = L
        IE(1) = 4
        IE(8) = IE(1)
        IE(2) = 14
        IE(7) = IE(2)
        IE(3) = 32
        IE(6) = IE(3)
        IE(4) = 50
        IE(5) = IE(4)
        DO 8    J = 1, N
        SUM = 0.0
        DO 7    M = 1, L
        X = RANDOM (0)
 7      SUM = SUM + X
 8      XM(J) = SUM / EL
        CALL UNIDIS (N, K, XM, F, IC)
        CALL GOOFIT (K, F, IE, IC)
        CALL EXIT
        END
```

CHAPTER NINE

Goodness of Fit

9.1 The chi-square distribution

The evaluation of computer-generated empirical sampling distributions, such as the one summarized in Table 8.1, has been difficult due to the lack of a statistical test for the hypothesis that the obtained empirical distribution differs only by chance from its theoretical sampling distribution. A tabular comparison of the two distributions was arranged for each sampling experiment on means (Tables 8.1, 8.2, and 8.3), and a visual comparison indicated some resemblance of the empirical to the theoretical normal distributions. It was noteworthy that the resemblance became stronger as the sample size (number of observations underlying each mean) increased. These comparisons are unsatisfying because there are no standards for deciding whether the discrepancies are due to sampling error or whether the discrepancies are so large as to justify the suspicion that some other form of population distribution is actually involved. After all, the theoretical normal distribution in the case of the means is only a hypothesis, which may be incorrect. Certainly the results for means based on sample size 10 look suspicious (Table 8.1 and Figure 8.1). The central-limit theorem asserts that the sampling distribution of means tends to normality as sample size increases, but does not say how large a sample is large enough. It has been suggested that about 30 is a sufficiently large N to warrant trust in the central-limit theorem's applicability, but on the other hand the random normal deviate generator RANDEV sums only 12 random fractions in the creation of a deviate. An adequate test of the normality of the generated deviates is also needed. In fact, since a major burden of this text is to convey the notion of sampling distributions for various statistics, and to develop confidence in them, a test for *goodness of fit* of an empirical or obtained distribution to a theoretical or expected distribution is most urgently needed.

The rectangular, the binomial, and the normal distributions have been discussed as examples of theoretical distributions. In order to present a goodness of fit test the authors must introduce yet another distribution, the χ^2 (chi-square) family of curves. You will never escape chi-square as long as you remain interested in statistics, and probably no worthy approach to the

THE CHI-SQUARE DISTRIBUTION

subject could get you through the first semester of study without an introduction to this widely used distribution. Unfortunately, χ^2 theory is as elusive as its applications are pervasive. Although your mastery of the χ^2 methodology will wait on further study in later semesters, some of the uses statisticians make of this distribution can at least be introduced at this time.

Perhaps the first point to be made about χ^2 is that it is the name for a *family* of curves. That is, there are many χ^2 distributions, so that it is necessary to employ a tag to indicate which one is involved in any specific application. This tag is an integer greater than zero, and is called the *number of degrees of freedom* (n.d.f. for short). Sometimes statisticians speak of "χ^2 with n.d.f. k" where k is an integer, and other times they indicate the n.d.f. by means of a subscript, "χ_k^2."

The best place to start an exposition of χ^2 is with consideration of χ_1^2. Chi-square with 1 degree of freedom is simply a squared random normal deviate. Thus

$$\chi_1^2 = z^2 = \frac{(X - \mu)^2}{\sigma^2} \qquad (9.1.1)$$

Note that while half the distribution of z is negative, the entire χ^2 distribution must be positive. The treatment of any $(X - \mu)$ and its counterpart with reversed sign $-(X - \mu)$ as equivalent establishes immediately that tabled values of χ^2 must be for two-tailed tests, as indeed they are. Since the bulk of z scores lie between -1 and $+1$ (68 percent of them, remember), the bulk of χ_1^2 values must lie between 0 and 1. Since a z smaller than -3 or greater than $+3$ is very rare, most of the range of χ_1^2 must fall between 0 and 9. However, like the normal curve, any positive value of χ^2, however large, has some probability of occurring. From Table 9.1 it can be seen that there is one chance in a thousand of obtaining $\chi_1^2 \geq 10.8$. Figure 9.1 shows the approximate form of the distribution curve for χ_1^2. For χ_1^2 the mean is 1 and the variance is 2.

Whereas χ_1^2 is the distribution of a squared random normal deviate, χ_2^2 is the distribution of the sum of two independent squared random normal deviates.

$$\chi_2^2 = \frac{(X_1 - \mu)^2}{\sigma^2} + \frac{(X_2 - \mu)^2}{\sigma^2} \qquad (9.1.2)$$

Its mean is 2 and its variance is 4. In general, χ_k^2 is the sum of k independent squared random normal deviates, and has mean k and variance $2k$.

$$\chi_k^2 = \sum_{i=1}^{k} z_i^2 = \sum_{i=1}^{k} \frac{(X_i - \mu)^2}{\sigma^2} \qquad (9.1.3)$$

Figure 9.1 also shows the approximate form of χ_2^2 and χ_8^2. The chi-square distribution is positively skewed for small k, but the skew is reduced as k increases. Chi-squares with n.d.f. of about 30 or more are approximately

Table 9.1 Cumulative Probabilities for χ^2 [a]

n.d.f. P:	.01	.05	.10	.90	.95	.99	.999
1	.0002	.004	.016	2.71	3.84	6.63	10.8
2	.020	.103	.211	4.61	5.99	9.21	13.8
3	.115	.352	.584	6.25	7.81	11.3	16.3
4	.297	.711	1.06	7.78	9.49	13.3	18.5
5	.554	1.15	1.61	9.24	11.1	15.1	20.5
6	.872	1.64	2.20	10.6	12.6	16.8	22.5
7	1.24	2.17	2.83	12.0	14.1	18.5	24.3
8	1.65	2.73	3.49	13.4	15.5	20.1	26.1
9	2.09	3.33	4.17	14.7	16.9	21.7	27.9
10	2.56	3.94	4.87	16.0	18.3	23.2	29.6
11	3.05	4.57	5.58	17.3	19.7	24.7	31.3
12	3.57	5.23	6.30	18.5	21.0	26.2	32.9
13	4.11	5.89	7.04	19.8	22.4	27.7	34.5
14	4.66	6.57	7.79	21.1	23.7	29.1	36.1
15	5.23	7.26	8.55	22.3	25.0	30.6	37.7
16	5.81	7.96	9.31	23.5	26.3	32.0	39.3
17	6.41	8.67	10.1	24.8	27.6	33.4	40.8
18	7.01	9.39	10.9	26.0	28.9	34.8	42.3
19	7.63	10.1	11.7	27.2	30.1	36.2	43.8
20	8.26	10.9	12.4	28.4	31.4	37.6	45.3
22	9.54	12.3	14.0	30.8	33.9	40.3	48.3
24	10.9	13.8	15.7	33.2	36.4	43.0	51.2
26	12.2	15.4	17.3	35.6	38.9	45.6	54.1
28	13.6	16.9	18.9	37.9	41.3	48.3	56.9
30	15.0	18.5	20.6	40.3	43.8	50.9	59.7
40	22.2	26.5	29.1	51.8	55.8	63.7	73.4
60	37.5	43.2	46.5	74.4	79.1	88.4	99.6
80	53.5	60.4	64.3	96.6	102.	112.	125.
100	70.1	77.9	82.4	118.	124.	136.	149.

[a] Values selected from Abramowitz and Stegun (1964), pp. 984–985.

normal in distribution, another result of the central-limit theorem. But note that the χ^2 distribution depends directly on the assumption of a sum of squares of independent *normal* variables each with mean 0 and variance 1. Statisticians speak of the normal curve as the *parent* distribution for chi-square.

Expression 9.1.3 defines χ_k^2 as a theoretical random variable. The entries in Table 9.1 stem from the cumulative probability function for that random variable. It is interesting to speculate that if RANDEV is generating random normal deviates, it can easily be employed to generate an empirical sampling distribution for χ_k^2, by a program which will call on RANDEV for k independent deviates and sum their squares, repeating this operation as many times as is desired. List 9.1 lists such a program. You should run it in your

THE CHI-SQUARE DISTRIBUTION

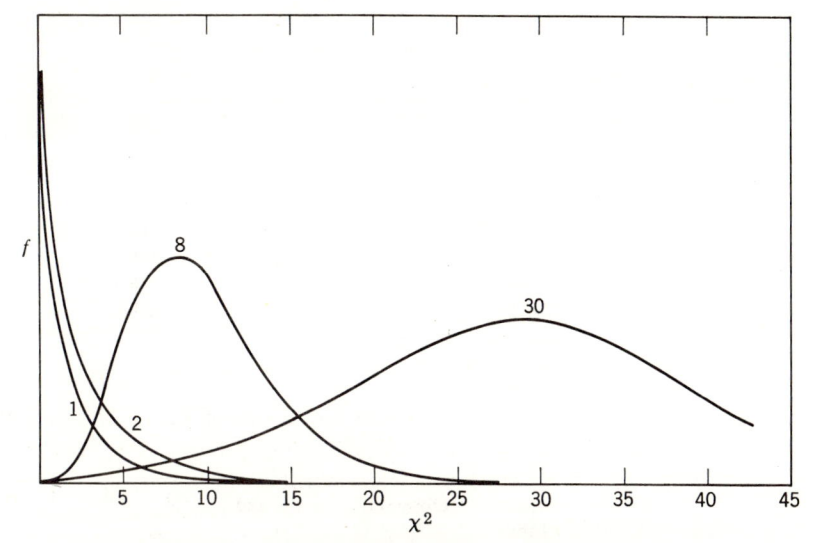

Fig. 9.1 Schematic distributions of chi-square with 1, 2, 8, and 30 degrees of freedom.

lab and see how the upper percentage points agree with Table 9.1. Of course, the outcome of this numbers game will be more predictable after, and if, we succeed in establishing credibility for the hypotheses that (1) RANDOM produces uniform random fractions, and (2) RANDEV produces from them normalized random deviates. Tables 9.2, 9.3, 9.4, and 9.5 report the author's results with List 9.1, each table using a different number of degrees of freedom. (The last column and the chi-square computation in each of these four tables will be explained in section 9.2).

Table 9.2 Empirical χ_2^2 from List 9.1 and Goodness of Fit to Theoretical χ_2^2

Class	Lower Limit	Upper Limit	Observed f	Expected f	$(O - E)^2 / E$
1	.000	.102	7	5	.800
2	.103	.210	5	5	.000
3	.211	.574	14	15	.067
4	.575	1.385	25	25	.000
5	1.386	2.772	26	25	.040
6	2.773	4.604	13	15	.267
7	4.605	5.990	3	5	.800
8	5.991	9.000	7	5	.800

$\chi_7^2 = 2.77$, $.05 < P < .10$

Table 9.3 Empirical χ_4^2 from List 9.1 and Goodness of Fit to Theoretical χ_4^2

Class	Lower Limit	Upper Limit	Observed f	Expected f	$\dfrac{(O-E)^2}{E}$
1	.000	.711	6	5	.200
2	.711	1.063	4	5	.200
3	1.064	1.922	17	15	.267
4	1.923	3.356	20	25	1.000
5	3.357	5.384	24	25	.040
6	5.385	7.778	15	15	.000
7	7.779	9.487	7	5	.800
8	9.488	20.000	7	5	.800

$\chi_7^2 = 3.31$, $.10 < P < .90$

Table 9.4 Empirical χ_8^2 from List 9.1 and Goodness of Fit to Theoretical χ_8^2

Class	Lower Limit	Upper Limit	Observed f	Expected f	$\dfrac{(O-E)^2}{E}$
1	.000	2.732	2	5	1.800
2	2.733	3.489	8	5	1.800
3	3.490	5.070	17	15	.267
4	5.071	7.343	33	25	2.560
5	7.344	10.21	24	25	.040
6	10.22	13.35	7	15	4.267
7	13.36	15.50	2	5	1.80
8	15.51	40.00	7	5	.800

$\chi_7^2 = 13.33$, $.90 < P < .95$

Table 9.5 Empirical χ_{30}^2 from List 9.1 and Goodness of Fit to Theoretical χ_{30}^2

Class	Lower Limit	Upper Limit	Observed f	Expected f	$\dfrac{(O-E)^2}{E}$
1	.000	18.48	5	5	.000
2	18.49	20.59	6	5	.200
3	20.60	24.47	9	15	2.400
4	24.48	29.33	32	25	1.960
5	29.34	34.79	25	25	.000
6	34.80	40.25	14	25	.067
7	40.26	43.76	6	5	.200
8	43.77	80.00	3	5	.800

$\chi_7^2 = 5.63$, $.10 < P < .90$

9.2 The goodness of fit test

Our test of how well an observed distribution agrees with a theoretical distribution, called a "goodness of fit" test, depends on a result originally established by Karl Pearson. He showed that under certain conditions a statistic can be computed from the comparison of two distributions which distributes approximately as χ^2. The first condition is that one of the distributions be viewed as a random sample from the other. The sample distribution is called the "observed" one, and the population distribution is called the "expected" one. Secondly, the values in both distributions must be grouped the same way into k, mutually exclusive and exhaustive categories, or class intervals. Thirdly, the total number of observations in the sample distribution N, must be very large. The second condition forces us to treat a continuous variable as if it were a discrete one. We are required to break the score range up into categories or classes, and to count the frequency of observations in each interval. Thus a continuous measure is reduced to a set of ordered classes, and in fact the goodness of fit test does not use even the order information, since the χ^2 computed would be the same for any reordering of the classes. The definition of these classes cannot be done too arbitrarily since it does influence the finding. The requirement of large N is here translated into the rule of thumb that the *expected* frequency in every class is to be at least 5. The authors' point of view is that the χ^2 goodness of fit test is an imperfect instrument to be used with caution, but one which can properly influence our convictions regarding the probable sources of sample distributions.

The computations for the goodness of fit test are quite straightforward. Class intervals are chosen, the theory is consulted to establish the probability of a random observation's falling in each class interval (p_i for the ith class), the sample size is chosen (N), and the expected frequencies are computed as

$$E_i = Np_i \qquad (9.2.1)$$

These are checked to assure that for any i, $E_i \geq 5$, and that $\sum_i E_i = N$. The sample data are then grouped into the classes, and the observed frequency in each class is counted (O_i). Chi-square can then be computed as

$$\chi_{k-m} = \sum_{i=1}^{k} \left[\frac{(O_i - E_i)^2}{E_i} \right] \qquad (9.2.2)$$

The number of degrees of freedom for this Pearson χ^2 is a little tricky, however. N.d.f. is the number of classes (k) minus the number of parameters fitted in determining the expected frequencies (m). This subtrahend is always at least 1, to account for forcing the sum of E_i to equal N, the arbitrarily

Table 9.6 Student's 100 Digits

1	7	7	0	1	4	4	3	2	0
1	8	8	0	5	4	4	1	3	2
9	8	8	6	4	2	1	5	5	7
0	6	4	4	3	2	1	3	1	0
2	5	4	3	3	5	4	3	2	5
7	7	7	5	7	6	6	3	3	1
0	2	0	2	0	2	4	4	5	4
2	5	4	5	3	2	6	7	8	9
0	2	5	4	0	2	3	2	1	0
9	9	8	7	9	9	8	7	9	9

selected sample size. Often the E_i will be forced to match the sample distribution in mean and variance as well, so that m will be 3.

To contrive an example of modest dimensions, let us suppose that a student in the class is unimpressed with FUNCTION RANDOM (J). "Well," he says, "I'm sure I can generate random digits just as effectively by simply calling them out, without recourse to multiplication modulo 1." The instructor having no choice but to accept the challenge, the contest is on. The student calls out 100 digits, watching carefully as the instructor writes them on the blackboard, one row at a time. The result is displayed in Table 9.6. The instructor then works through the steps of the goodness of fit test.

1. Choose class intervals.
   He arbitrarily defines five classes: 0–1, 2–3, 4–5, 6–7, 8–9.
2. Establish probabilities for classes.
   Since random digits should be uniformly distributed, the probability that a given digit falls in class i is .20 for all five classes.
3. Select a sample size.
   He has already established that $N = 100$.
4. Compute expected frequencies for the classes.
   Since $E_i = Np_i$, and p_i is constant for all classes, $E_i = 20$ for all i.
5. Group the sample data into the classes and tabulate the frequency in each class.
   He makes the following tabulation:

   ```
   0–1   ||||  ||||  ||||  ||||
   2–3   ||||  ||||  ||||  ||||  ||||
   4–5   ||||  ||||  ||||  ||||  ||||
   6–7   ||||  ||||  ||||
   8–9   ||||  ||||  ||||
   ```

6. Compute the chi-square.
   He arranges Table 9.7, and sums the last column for χ^2: (He determines that there are 4 degrees of freedom by subtracting the number

THE GOODNESS OF FIT TEST

of parameters fitted—which is one, N, since he did not use the mean and variance for this particular uniform distribution—from the number of classes. He gets the probability statement by reference to Table 9.1)

The instructor, with a red face, certifies the victorious student as a bona fide random digits generator? Certainly not. He argues, cogently if belatedly, that while a random digits generator should produce a reasonable approximation to a uniform distribution over a large number of trials, we cannot afford to believe that just any approximately uniform array of numbers is

Table 9.7 Chi-Square Analysis of Student's 100 Digits

Class	Expected	Observed	$(O - E)^2 / E$
0–1	20	20	0
2–3	20	25	1.25
4–5	20	25	1.25
6–7	20	15	1.25
8–9	20	15	1.25
	$\chi_4^2 = 5.0$,	$.10 < P < .90$	

random. In the present case, the digits are clearly not independent, because the student was reviewing the written record of what he had produced before as he generated the next digit. The last row he dictated obviously was designed to correct for an apparent underproduction of 7's, 8's, and 9's in the previous nine rows. A process with a memory cannot be a thoroughly random process. The moral is that while we may be encouraged if we can show by a goodness of fit test that RANDOM appears to be sampling a uniform distribution, this demonstration will not entirely prove the randomness of the samples generated.

It is instructive to consider the outcome of a χ^2 test on the same data, but using a different set of class intervals, which can easily be arranged if we take each digit as an interval itself, so that there are ten classes, p_i is uniformly .10 and E_i is uniformly 10 for all classes, and Table 9.8 results. The probability for this chi-square is similar, and we have similar confidence in our failure to reject the null hypothesis.

Suppose someone suggests that each of 100 students in the class write on a ballot a single digit, all writing simultaneously and independently, as an attempt at random digits generating. This appeals to the instructor, and the Table 9.9 data and analysis result from the experiment. On this evidence the procedure does seem to barely qualify as a random digits generator, yet the class seems to have "favorite" digits. Which digits were they?

Table 9.8 Chi-Square Analysis of Classroom Experiment

Class	Expected	Observed	$(O-E)^2/E$
1	10	11	.1
1	10	9	.1
2	10	14	1.6
3	10	11	.1
4	10	14	1.6
5	10	11	.1
6	10	5	2.5
7	10	10	.0
8	10	7	.9
9	10	8	.4

$\chi_9^2 = 7.4$, $.10 < P < .90$

Table 9.9 Chi-Square Analysis of Second Classroom Experiment

DATA TABLE

0	8	4	3	6	7	7	0	7	3
1	5	8	2	4	1	4	4	0	1
4	4	8	9	4	5	7	2	4	5
8	5	9	1	3	7	4	6	4	0
5	9	4	2	2	6	5	9	9	7
7	2	5	7	9	1	3	7	2	9
9	2	8	8	2	1	5	4	8	9
5	8	8	5	9	1	1	9	5	9
3	5	6	7	8	8	8	8	8	9
0	2	5	9	4	1	9	8	0	3

CHI-SQUARE ANALYSIS

Class	Expected	Observed	$(O-E)^2/E$
0	10	7	.9
1	10	6	1.6
2	10	11	.1
3	10	6	1.6
4	10	12	.4
5	10	14	1.6
6	10	4	3.6
7	10	10	.0
8	10	15	2.5
9	10	15	2.5

$\chi_9^2 = 14.8$, $.90 < P < .95$

9.3 Applications to Monte Carlo results

The laboratory experiments for the earlier chapters have left the class with computer printouts containing empirical sampling distributions from three ostensible sources: (1) the uniform distribution within the range $.0000 \leq x \leq .9999$, for which $\mu = .5000$ and $\sigma^2 = .0833$, sampled 500 times by UNIFORMITY OF RANDOM FRACTIONS EXPERIMENT (List 5.2) (review Figure 5.1 and Table 5.1); (2) the unit normal distribution, $N(0, 1)$, sampled 1000 times by TEST FUNCTION RANDEV (O) (List 7.2) (review Figures 7.11 and 7.12, and Table 7.4); and (3) the distribution of sample means, hypothetically $N(\mu_x, \sigma_x/\sqrt{N})$. Note that the third source has been sampled three times, with three different values for N, the number of observations underlying each sample mean. From MONTE CARLO STUDY OF THE SAMPLE VARIANCE (List 5.3) a sample of 200 means based on $N = 10$ was obtained (review Figure 8.1 and Table 8.1). From MONTE CARLO ON THE SAMPLE MEAN (List 8.1) two additional samples of 200 means, one based on $N = 30$ and the other on $N = 60$, were obtained (review Figures 8.2 and 8.3, and Tables 8.2 and 8.3). For each of these five samples we have a hypothetical parent distribution, but we have had to be satisfied with subjective comparisons of the obtained distributions with the expected distributions. We have learned statistical tests for the obtained sampling distributions' means in Chapter 8, but we have had no statistical criterion for the comparison of the obtained sampling distributions' variances with those expected, and no criterion for the match of the complete obtained frequency distribution to the hypothesized parent distribution. Statisticians do have a test for single variances, which you will acquire in the next chapter. Interestingly, the test criterion distributes as chi-square. Right now we can apply the χ^2 goodness of fit test to our hypotheses regarding parent distributions for our empirical distributions.

Lazy arithmeticians that we are, we prefer to compute the required χ^2 tests on the computer. Fortunately, in our Monte Carlo programs SUBROUTINE UNIDIS (List 4.3) has already accomplished the grouping of the data into appropriate classes and the tabulating of class observed frequencies. The programs have then called on SUBROUTINE GOOFIT (List 9.3) to compute $(O - E)^2/E$ for each class, and sum to χ^2.

Before we go ahead to interpret our Monte Carlo goodness of fit results, however, we should review the steps by which we arrived at the expected frequencies for classes in each of the tables. For Table 9.10, based on List 5.2 outcomes, where the hypothesized population distribution is uniform, the task was to recognize that the equal sizes of the class intervals chosen implied equal class probabilities. In this case, $p_i = .10$ for every i, and $Np_i = (500)(.10) = 50$ established the expected frequency for each class.

Table 9.10 500 Empirical Random Fractions from List 5.2 and Goodness of Fit to Uniformity Hypothesis

Class	Lower Limits	Upper Limits	(O) Observed f	(E) Expected f	$(O-E)^2/E$
1	.000	.099	43	50	.980
2	.100	.199	44	50	.720
3	.200	.299	53	50	.180
4	.300	.399	68	50	6.480
5	.400	.499	50	50	.000
6	.500	.599	54	50	.320
7	.600	.699	52	50	.080
8	.700	.799	57	50	.980
9	.800	.899	37	50	3.380
10	.900	1.000	42	50	1.280

$\chi_9^2 = 14.40, \quad P \sim .90$

For the 1000 random deviates, as displayed in Table 9.13, the method was to set up uniform intervals and then consult a table of the unit normal curve for class probabilities. The smaller expected frequencies for the tails of the curve can be justified because we have more interest in the goodness of fit in the tails of the normal curve, where critical statistical inferences are made, than in the center. We really shouldn't use cells with $E = 1$, of course, but we want to show the good fit obtained in the extreme tails.

Table 9.11 100 Empirical Random Fractions from List 5.2 and Goodness of Fit to Uniformity Hypothesis

Class	Lower Limit	Upper Limit	(O) Observed f	(E) Expected f	$(O-E)^2/E$
1	.000	.099	8	10	.400
2	.100	.199	8	10	.400
3	.200	.299	13	10	.900
4	.300	.399	8	10	.400
5	.400	.499	9	10	.100
6	.500	.599	9	10	.100
7	.600	.699	11	10	.100
8	.700	.799	8	10	.400
9	.800	.899	15	10	2.500
10	.900	1.000	11	10	.100

$\chi_9^2 = 5.40, \quad .10 < P < .90$

APPLICATIONS TO MONTE CARLO RESULTS

A unit normal curve table does not give probabilities for z score intervals directly. It is necessary to work from the cumulative probabilities, by subtracting the cumulative probability for the lower bound of the interval from the cumulative probability for the upper bound. That is, if z_L and z_U are the lower and upper bounds for the z score interval:

$$p_i = P(z_{U_i}) - P(z_{L_i}) \qquad (9.3.1)$$

For example, for the class interval $+1.00$ through $+2.00$ the unit normal curve table reveals that $P(z = 1.00) = .841$ and $P(z = 2.00) = .977$. Then

$$p_i = P(2.00) - P(1.00) = .977 - .841 = .136$$

There are two approaches to selecting class intervals for a grouped-data frequency distribution and goodness of fit testing. In one approach uneven class intervals are structured to yield equal sized theoretic frequencies (except in the tails, where more detail is desired). The upper half of Table 9.12 provides a useful set of unevenly spaced z values with equal increments

Table 9.12 Twelve-Class Unit Normal Distribution

Interval	Lower Limit (z)	Class Probability	Cumulative Probability
PART ONE: SYMMETRIC CLASS PROBABILITIES			
1	$-\infty$	.05	.05
2	-1.64	.05	.10
3	-1.28	.10	.20
4	$-.84$	.10	.30
5	$-.52$	.10	.40
6	$-.25$	.10	.50
7	.00	.10	.60
8	.25	.10	.70
9	.52	.10	.80
10	.84	.10	.90
11	1.28	.05	.95
12	1.64	.05	1.00
PART TWO: SYMMETRIC CLASS INTERVALS			
1	$-\infty$	.005	.005
2	-2.5	.020	.025
3	-2.0	.040	.065
4	-1.5	.095	.160
5	-1.0	.150	.310
6	$-.5$	.190	.500
7	0.0	.190	.690
8	.5	.150	.840
9	1.0	.095	.935
10	1.5	.040	.975
11	2.0	.020	.995
12	2.5	.005	1.000

in probability corresponding. The alternative approach is to employ equally spaced z values with corresponding uneven theoretic frequencies. Tables 8.1, 8.2, and 8.3 exemplify this approach. The lower half of Table 9.12 provides a useful set of evenly spaced z values with unequal increments in probability. The choice between these two approaches, each of which is perfectly valid, is an example of the opportunities for expression of "research style" in the organization of data analyses. You can do all the testing of goodness of fit to the normal curve you want using the one set of class intervals and expected probabilities recorded in Table 9.12, if you first transform to your observed scores units from the z score distribution, by multiplying the z score by the observed standard deviation, and adding the observed mean. If your N is too small to support this many intervals, just pool some adjacent intervals. Remember that n.d.f. for χ^2 is three less than the number of intervals employed when you fit to the sample mean and standard deviation. List 9.2 provides a program for running goodness of fit tests.

Table 9.10 reports the authors' goodness of fit test for 500 fractions generated by List 5.2 to the rectangular distribution expected. Note that the obtained chi-square does not disqualify this as a random sample from a uniform source, for we cannot reject at the .05 level. For a second sample of 100 fractions from List 5.2, the chi-square falls in a nice, safe region, as indicated by Table 9.11.

The goodness of fit test for 1000 random normal deviates generated by List 7.2 is reported in Table 9.13. Again the probability of the test statistic is in a safe region and we do not reject the null hypothesis.

Since the expected frequencies were arrayed alongside the observed frequencies in Tables 8.1, 8.2, and 8.3, new tables are not included in this

Table 9.13 1000 Empirical Random Normal Deviates Generated by List 7.2 and Goodness of Fit to $N(0, 1)$

Class	Lower Limit	(O) Observed f	(E) Expected f	$(O - E)^2 / E$
1	$-\infty$	0	1	1.000
2	-3.00	23	22	0.045
3	-2.00	146	136	0.735
4	-1.00	316	341	1.833
5	0.00	343	341	0.012
6	1.00	150	136	1.441
7	2.00	21	22	0.045
8	3.00	1	1	0.000

$\chi_7^2 = 5.1, \quad .10 < P < .90$

chapter. However, for Table 8.1 the test result is $\chi_7^2 = 11.5$, $.10 < P < .90$, so that we cannot reject the null hypothesis of a normal parent distribution for this sample of 200 means based on 10 random fractions each. Actually, we do not expect the central-limit theorem to fit the means of such small samples to the normal distribution very well, but our test of goodness of fit is not very powerful, either. That is, departures from the expected distribution must be rather strong to lead to a rejection. For Table 8.2 the test result is $\chi_7^2 = 8.5$, $.10 < P < .90$, indicating that the authors' sample of 200 means of 30 random fractions fits the normal hypothesis well enough, and for Table 8.3 the test result is $\chi_7^2 = 7.4$, $.10 < P < .90$, indicating a reasonable fit to normality for 200 means of 60 random fractions.

These results build some confidence in our random fractions generator and our random normal deviates generator. Look now at the test results for Tables 9.2, 9.3, 9.4, and 9.5. Note that our samples of 100 chi-squares with 2, 4, 8, and 30 degrees of freedom all qualify. Perform similar analyses on your class results from List 9.1 and compare with these outcomes.

9.4 Summary

Chi-square with k degrees of freedom is a theoretical distribution which may be analyzed as the sum of k independent squared random normal deviates. It has mean k and variance $2k$. For small k, χ_k^2 has a positively skewed distribution, but for k larger than 30 the distribution is approximately normal, as the central-limit theorem would suggest.

FORTRAN LIST 9.1 can generate N points in an empirical chi-square distribution as the sum of M independent squared random normal deviates, taking the deviates from FUNCTION RANDEV (K).

Chi-square has many uses in applied statistics, but the special use in Pearson's *goodness of fit test* is of strategic importance in this text, where numerous empirical distributions require testing for goodness of fit to their hypothetical parent distributions.

FORTRAN LIST 9.2 reads in the observed and expected frequencies for classes in a grouped data frequency distribution and computes a goodness of fit chi-square. The program reports the chi-square contribution of each cell as well as the total chi-square, making it easy to spot the cells in which the largest discrepancies from theory occur.

Getting the expected frequencies for the classes is the most difficult part of a goodness of fit test. The principle is to allow the distribution theory to dictate the allocation of frequencies to classes, given the total frequency and the parameters or their sample estimates, usually the mean and the standard deviation. The authors have provided expected frequencies for appropriate classes for all the required applications. When population parameters are used, n.d.f. $= k - 1$, whereas when sample estimates of the mean and the standard deviation are used, n.d.f. $= k - 3$, where k is the number of class intervals.

EXERCISES AND QUESTIONS

1. Can you think of another check on randomness which the instructor might have used with those numbers generated by the student in Table 9.6? What about frequency with which digits are immediately repeated? How often would you expect that to happen in a table of 100 random digits?
2. Do the lab exercise outlined at the end of Section 9.1 using List 9.1.
3. Also be sure to interpret your List 9.1 results. Since you used a different starting random fraction, how do your results compare with those presented here? How do they compare with theory?

```
C        LIST 9.1   CHI-SQUARE GENERATION EXPERIMENT
C
C        THIS PROGRAM GENERATES AND DISTRIBUTES 100 RANDOM POINTS EACH FROM
C        CHI-SQUARES WITH 2,  4,  8,  AND  30  DEGREES OF FREEDOM.
C        REQUIRED SUBROUTINES ARE RANDOM, RANDEV, UNIDIS, AND GOOFIT.
C        THE SINGLE DATA CARD CONTAINS AN 8-DIGIT STARTING NUMBER FOR
C            RANDOM IN COLUMNS 1-8.
C
         DIMENSION   C(1000),  F(100),  IC(100),  IE(100)
C
         N = 100
         K = 8
         IE(1) = 5
         IE(2) = IE(1)
         IE(7) = IE(1)
         IE(8) = IE(1)
         IE(3) = 15
         IE(6) = IE(3)
         IE(4) = 25
         IE(5) = IE(4)
         Y = RANDOM (1)
         M = 2
         WRITE (6,1)
1        FORMAT (1H1)
         WRITE (6,2)     M
2        FORMAT (40H0100 POINTS FROM CHI-SQUARE WITH NDF =     I3)
         DO 3   J = 1, N
         C(J) = 0.0
         DO 3   L = 1, M
         Z = RANDEV (0)
3        C(J) = C(J) + Z * Z
         F(1) = .000
         F(2) = .103
         F(3) = .211
         F(4) = .575
         F(5) = 1.386
         F(6) = 2.773
         F(7) = 4.605
         F(8) = 5.991
         CALL UNIDIS (N, K, C, F, IC)
         CALL GOOFIT (K, F, IE, IC)
```

SUMMARY

```
C
      M = 4
      WRITE (6,2)     M
      DO 4    J = 1, N
      C(J) = 0.0
      DO 4    L = 1, M
      Z = RANDEV (0)
    4 C(J) = C(J) + Z * Z
      F(2) = .711
      F(3) = 1.064
      F(4) = 1.923
      F(5) = 3.357
      F(6) = 5.385
      F(7) = 7.779
      F(8) = 9.488
      CALL UNIDIS (N, K, C, F, IC)
      CALL GOOFIT (K, F, IE, IC)
C
      M = 8
      WRITE (6,2)     M
      DO 5    J = 1, N
      C(J) = 0.0
      DO 5    L = 1, M
      Z = RANDEV (0)
    5 C(J) = C(J) + Z * Z
      F(2) = 2.733
      F(3) = 3.490
      F(4) = 5.071
      F(5) = 7.344
      F(6) = 10.22
      F(7) = 13.36
      F(8) = 15.51
      CALL UNIDIS (N, K, C, F, IC)
      CALL GOOFIT (K, F, IE, IC)
C
      M = 30
      WRITE (6,2)     M
      DO 6    J = 1, N
      C(J) = 0.0
      DO 6    L = 1, M
      Z = RANDEV (0)
    6 C(J) = C(J) + Z * Z
      F(2) = 18.49
      F(3) = 20.60
      F(4) = 24.48
      F(5) = 29.34
      F(6) = 34.80
      F(7) = 40.26
      F(8) = 43.77
      CALL UNIDIS (N, K, C, F, IC)
      CALL GOOFIT (K, F, IE, IC)
      CALL EXIT
      END
```

List 9.1. (*continued*)

```
C     LIST 9.2    GOODNESS OF FIT
C
C     THIS PROGRAM COMPUTES A GOODNESS OF FIT CHI-SQUARE FROM INPUT CONSISTING
C     OF THE OBSERVED AND EXPECTED FREQUENCIES FOR DATA GROUPED INTO N CLASSES.
C
C     THE FIRST DATA CARD CONTAINS THE PROBLEM NUMBER (COLS 1-5) AND N (COLS 6-8).
C     EACH ADDITIONAL DATA CARD CONTAINS THE LOWER BOUND OF THE CLASS INTERVAL
C     (COLS 2-5), THE UPPER BOUND OF THE CLASS INTERVAL (COLS 6-10), THE OBSERVED
C     FREQUENCY FOR THE CLASS (COLS 11-15), AND THE EXPECTED FREQUENCY FOR THE
C     CLASS (COLS 16-20).
C
      READ (5,2)   NPROB,  N
   2  FORMAT (I5, I3)
      WRITE(6,3)   NPROB,  N
      WRITE(7,3)   NPROB,  N
   3  FORMAT(11H1PROB. NO. I6,31H.  GOODNESS OF FIT FOR DATA IN I4,8H CL
     CASSES)
      WRITE(6,4)
      WRITE(7,4)
   4  FORMAT(51H0CLASS LOWER   UPPER   OBSERVED   EXPECTED    (O-E)**2/E)
C
      CHISQ = 0.0
      DO 10  J = 1, N
      READ (5,5)   XLOER,  UPPER,  OBSER,   EXPEC
   5  FORMAT (4F5.0)
      CHI = ((OBSER - EXPEC)**2) / EXPEC
      CHISQ = CHISQ + CHI
      WRITE(6,6)    J,  XLOER,  UPPER,  OBSER,  EXPEC,  CHI
  10  WRITE(7,6)    J,  XLOER,  UPPER,  OBSER,  EXPEC,  CHI
   6  FORMAT (1H0, I5, F7.3, 1X, F7.3, 3X, F6.0, 4X, F6.0, 4X, F7.3)
C
      WRITE(6,7)    CHISQ
      WRITE(7,7)    CHISQ
   7  FORMAT (14H0CHI-SQUARE = F8.3)
C
      CALL EXIT
      END


C     LIST 9.3    SUBROUTINE GOOFIT
C
C     THIS SUBROUTINE COMPUTES A GOODNESS OF FIT CHI-SQUARE.
C     THE ARGUMENTS ARE
C         K      NUMBER OF CLASSES IN GROUPED DATA FREQUENCY DISTRIBUTION
C         F      VECTOR OF LOWER LIMITS OF CLASSES
C         IE     VECTOR OF EXPECTED FREQUENCIES OF CLASSES
C         IC     VECTOR OF OBSERVED FREQUENCIES OF CLASSES.
C     VALUES FOR ALL ARGUMENTS ARE SUPPLIED FROM MAIN PROGRAM.
C
      SUBROUTINE GOOFIT (K,  F,  IE,   IC)
C
      DIMENSION   F(100),   IE(100),   IC(100)
C
      WRITE (6,1)    K
   1  FORMAT (31H0  GOODNESS OF FIT FOR DATA IN    I4,8H CLASSES  )
      WRITE (6,2)
   2  FORMAT (53H0CLASS LOWER LIMIT    OBSERVED    EXPECTED    (O-E)**2/E )
      CHISQ = 0.0
C
      DO 3   J = 1, K
      OBSER = IC(J)
      EXPEC = IE(J)
      CHI = ((OBSER - EXPEC)**2) / EXPEC
      CHISQ = CHISQ + CHI
   3  WRITE (6,4)     J, F(J),  IC(J),  IE(J),   CHI
   4  FORMAT (1H0,I5,F10.4,6X,I6,5X,I6,F12.3)
C
      NDF = K - 1
      WRITE (6,5)     NDF,  CHISQ
   5  FORMAT (21H0CHI-SQUARE WITH NDF    I5,  4H = F10.3)
      RETURN
      END
```

CHAPTER TEN

Other Uses of Chi-Square Distribution

10.1 Testing a hypothesis about a variance

As we saw in Chapter 8, there is a great deal of interest in testing hypotheses involving means. Researchers are often also interested in comparing variability. It is possible, for example, that a particular experimental treatment produced increased individual differences. If an investigator drew a random sample from a population of known variance, σ_X^2, and then wished to compare the variance in his sample after treatment with the population variance, he would need to know the sampling distribution of the variance in order to decide if his posttreatment variance, $\hat{s}_X^2$, was unusually large.

So far, you have received no indication of the form of the sampling distribution for a variance. If pressed for a hypothesis at this point, we trust that you would guess that sample variances distribute normally. This is not a bad guess, because for variances computed from very large random samples of normal populations the sampling distribution does tend to normality. For variances computed from moderate or small samples of normal populations of scores, however, the sampling distribution is disturbed by the inevitable errors in estimating the population mean. We have seen that dividing the sum of squared deviations from a sample mean by $N - 1$ yields an *unbiased* estimate of the variance, but this only implies that the sampling distribution of the unbiased estimate, $\hat{s}_X^2$, centers over the population variance, σ_X^2. The correction centers but does not normalize the sampling distribution.

For random samples of size N from a normal population of scores the following statistic distributes approximately as χ^2 with $N - 1$ degrees of freedom:

$$\chi^2_{N-1} = \frac{(N-1)\hat{s}_X^2}{\sigma_X^2} \tag{10.1.1}$$

This statistic can be used to test the null hypothesis that a random sample from a *population known to be normal* is from a population with variance σ_X^2, so that the observed discrepancy $\hat{s}_X^2 - \sigma_X^2$ is due to sampling error.

The central-limit theorem assures that the sampling distribution of means based on large samples of scores is approximately normal. Therefore, it should be an appropriate application of the above test if we ask whether $s_M^2 - \sigma_M^2$ is attributable to chance, where s_M^2 is the observed unbiased estimate of the known variance error of the mean for a given sample size, $\sigma_M^2 = \sigma_X^2/N$.

In our MONTE CARLO STUDY OF THE SAMPLE MEAN of Section 8.1 we started with a random fraction .00000101 and $N = 30$, and obtained a sampling variance of the mean $s_M^2 = .00275$. Since $\sigma_X^2 = .0833$ for the uniform fractions sampled, $\sigma_M^2 = .0833/30 = .00278$. We now want to ask whether the difference between σ_M^2 and s_M^2 can be attributable to sampling error. Using Equation 10.1.1,

$$\chi_{29}^2 = \frac{(29)(.00275)}{.00278} = 28.7$$

The probability of a χ_{29}^2 as large as 28.7 or larger is about .50, so the null hypothesis is not rejected.

The same program reported for $N = 60$, $s_M^2 = .00128$. For this sampling distribution, $\sigma_M^2 = .0833/60 = .00139$

$$\chi_{59}^2 = \frac{(59)(.00128)}{.00139} = 54.3$$

The probability of a chi-square as large or larger than this is again about .50 and again the null hypothesis is not rejected. There is no reason in the sample variances of these means to dispute the notion that they are samples from normally distributed populations of means.

10.2 The significance of a difference between proportions

We have looked at statistical inference about a mean and a difference between means, about the parentage of a data distribution, and about a variance. All these examples have involved only one measurement variable, X, which we have called the empirical or observed distribution. Even the test for a difference in means, while comparing two estimates from two samples, assumes the same variable sampled twice, either by two groups of subjects (independent means), or by the same group observed twice (dependent means). Before considering the logic of inference in greater detail, we want to equip you with your first example of inference involving the relationship between two empirical distributions. The form of the question in this example is perhaps the single most pervasive semantic form in applied statistics, regardless of the field of application. The general situation is that two different measurement procedures have been applied to one group of subjects, where the subjects represent a random sample from a population of interest.

The general question is: Do the data warrant an assumption of an association or dependency between the two variables in the population?

Please notice at once that the question is *not* whether variable X *causes* variable Y, or vice versa. Statistical inference on bivariate relationships is not an inquiry into causality. The notions of dependency, association, and correlation assert only that two variables vary together. This may be due to one variable causing the variation in the other, or to the pair of variables having a common cause, or maybe because the two variables are associates in a complex system of variables which has some dynamics as a system. For example, a very strong positive relationship has existed over the years between the number of Methodist ministers in New England and the number of barrels of rum that have been shipped into the port of Boston. The more ministers there were, the more rum imported. The mere demonstration of the relationship does not allow you to say the ministers drank the rum, or were somehow responsible for rum consumption. Perhaps a third variable such as the population of New England, affected both the amount of rum and the number of ministers similarly! The point is that scientists very frequently look at two variables in a sample and ask whether there is sufficient evidence for assuming an association of the variables in the population. This consideration is prior to the consideration of the possible explanations of a real association. Even when the investigator has measured three or more variables on his subjects which he suspects are systematically interrelated, he will begin his probing of the system by looking at each bivariate relationship in his data first.

For the present, we are going to concern ourselves only with the question of the *existence* of a dependency, not with the more exciting question of the *strength* of an existing dependency. In Chapters 13 and 14 you will be introduced to the game of estimating the degree of dependency in a bivariate relationship, and then we expect you will become "hooked" on statistics permanently. The null hypothesis in the question of the existence of a relationship is, of course, that the two variables are independent in the population, and that whatever degree of association they show in the sample data is attributable to the operation of chance in random sampling. The immediate object is to confirm or reject this null hypothesis of no real relationship between X and Y.

Variables X and Y can have any of a variety of scaling properties in research, and need not have the same scaling properties. Both could be nominal, ordinal, or interval variables, or any mix could occur. Statistics provides a variety of models for coping with this variety of possibilities. Most of these you will have to study in later semesters, although the most useful of all, the Pearson correlation coefficient for a relationship between two ordinal, interval, or ratio variables, is presented in Chapter 14. Here we present the simplest situation, which concerns the possible association of two dichotomous nominal variables. On this type of variable every subject is classified into one of two exclusive and exhaustive categories. Sex is the most obvious dichotomous nominal variable. As a second example of a dichotomous nominal variable we shall take a questionnaire item response of the forced-choice type, where every respondent must answer *yes* or *no*.

The bivariate distribution for two such scales forms a four cell table, called a *contingency table*, like Table 10.1. If the categories of such a contingency table are mutually exclusive and exhaustive, *and* if each observation is independent of every other observation, *and* if the set of observations

Table 10.1 Sex by Response Distribution

	Response		
Sex	Yes	No	Totals
Male	30	20	50
Female	10	40	50
Totals	40	60	$N = 100$

are a large random sample from a population of potential observations, a statistic which distributes approximately as χ^2 can be computed from the data as a test criterion for the null hypothesis.

This section has been titled "the significance of a difference between proportions" because such data are frequently presented as a pair of proportions. For example, such a presentation would be:

Proportion of Each Sex Responding "Yes"
Male   .60
Female   .20

The null hypothesis could be stated as the proposition that the proportion of *yes* responses for males in the population is equal to the proportion of *yes* responses for females in the population. The test for equality of two independent proportions such as these is equivalent to the test for independence in a fourfold (four cell) contingency table. The authors prefer to see the data arrayed in a fourfold table in terms of observed frequencies because such a table tells the whole story of the observed bivariate distribution. They favor the presentation of a table of percentages or proportions *in addition to* the table of frequencies. Percentages and proportions alone have the unfortunate feature of concealing sample size!

The Pearson chi-square for a fourfold table has 1 degree of freedom.

This is because the marginal distributions (the row totals and the column totals) are taken as "givens" in the establishing of the expected frequencies for the cells. If the row and column totals are accepted as fixed entities, the observation of a frequency for any one of the four cells uses up the available degrees of freedom in the table. Placing a frequency in any one cell forces the other three cell frequencies to assume those unique values which will satisfy the fixed marginal distributions. In our example, the observation that 30 males responded *yes* uses up the 1 degree of freedom, given that there are 50 males and 50 females in the sample, and given that there are 40 *yes* responses and 60 *no* responses. If 30 males responded *yes*, the other 20 must have responded *no*. If 30 of the *yes* responses belonged to males, the other 10 *yes* responses must have belonged to females, and if 20 *no* responses belonged to males, the other 40 *no* responses must have belonged to females.

Taking the observed marginal distributions as givens may be disconcerting, but it is reasonable because the question under scrutiny concerns the possible association of the two variables in a population, not the characteristics of the distribution of either variable. Actually, if the sampling has been done by a simple random draw, as suggested, the observed marginal distributions themselves represent findings for the study. That is, there is the finding that the best estimate of the population sex ratio is 50/50, and the finding that the best estimate for the population response ratio is 40 *yes*/60 *no*. But these findings are not germane to the question of association. A more complicated sampling strategy, by which a fixed sample ratio for one of the variables, probably sex, was imposed prior to drawing the sample, could just as legitimately support the inquiry into association. The exact 50/50 split on sex in our example makes it look suspiciously like a *stratified random sample*, although we have imagined this as the unusual result of a simple random draw. If there had been stratification on sex, the 50/50 marginal distribution would be an arranged event, not a finding.

The expected frequency for any cell of a contingency table (E_{ij}) is a function of the row total for the row in which the cell occurs (R_i), and of the column total for the column in which the cell occurs (C_j), and of the total sample size (N). It represents the most likely cell frequency (if there is no relationship between the two variables) given these three parameters, and is computed as:

$$E_{ij} = \frac{R_i \cdot C_j}{N} \qquad (10.2.1)$$

The rule of thumb that no expected cell value is to be less than 10 applies to fourfold tables. Later you will learn how to make adjustments in situations where the expected frequencies are between 5 and 10. The full numerical roster for computing chi-square on a fourfold contingency table may be

Table 10.2

		$I \times J$ Contingency		
		\multicolumn{2}{c|}{J Effect}	Totals
		$J = 1$	$J = 2$	
I Effect	$I = 1$	O_{11} (E_{11})	O_{12} (E_{12})	R_1
	$I = 2$	O_{21} (E_{21})	O_{22} (E_{22})	R_2
Totals		C_1	C_2	N

represented symbolically as shown in Table 10.2. For the example, E_{11} (male, *yes*) is (50)(40)/100 or 20. Table 10.3 presents the expected values for the four cells, and the χ^2 addends. There is ample evidence against the null hypothesis. Apparently response tendency on this item is associated with sex of respondent.

This method of analyzing contingency tables can be extended to situations in which there are more than two categories for either or both of the variables. For example, Table 10.4 reports the amount of teaching experience for the teachers participating in the HOSC Instruction Project summarized in Chapter 1. Of the 106 teachers for whom years of teaching experience was known, 52 teachers were in the experimental group and 54 were control group teachers. Chi-square analysis of this 2 × 4 contingency table is a check to make sure that there was no relationship between amount of teaching experience and whether or not the teacher was in the experimental or control group.

Applying the formula 10.2.1 to the observed values at the top of Table 10.4 yields the table of expected frequencies at the bottom of the table. Then

Table 10.3 χ^2 Analysis for Sex by Response

Cell	Expected	Observed	$(O-E)^2/E$
Male, yes	20	30	5.00
Male, no	30	20	3.33
Female, yes	20	10	5.00
Female, no	30	40	3.33
	$\chi_1^2 = 16.66,$	$P > .999$	

Table 10.4 Teaching Experience of the HOSC Teachers

	Experimental	Control	Total
	OBSERVED		
5 years or less	9	17	26
6 to 10 years	16	11	27
11 to 20 years	15	12	27
More than 20 years	12	14	26
Total	52	54	106
	EXPECTED		
5 years or less	12.8	13.2	26
6 to 10 years	13.2	13.8	27
11 to 20 years	13.2	13.8	27
More than 20 years	12.8	13.2	26
Total	52	54	106
$\chi_3^2 = 3.84$,	$.10 < P < .90$		

applying $(O - E)^2/E$ to each cell and summing across all eight cells yields a χ^2 of 3.84. Given the row and column totals of Table 10.4, how many cell values could you define before the entire set of eight cell values is fixed? The answer is three, and that is the number of degrees of freedom for the chi-square [n.d.f. = (number of rows − 1)(number of columns − 1)].

Entering Table 9.1 with 3 degrees of freedom we see that obtaining a chi-square of 3.84 or smaller has a P between .10 and .90 so that we do not reject the null hypothesis. It is therefore safe to assume that the two groups of teachers, experimental and control, did not differ significantly on amount of teaching experience. Of course the teachers were randomly assigned to the two groups so it would be surprising to find a nonrandom difference between the two groups. Since it is not impossible, however, a check on key variables is not a waste of time, even though 95 percent of the time it will seem to be.

Incidentally, there are short-cut desk calculator methods which bypass the need to compute the expected frequencies, but we prefer to see the expected frequencies for purposes of interpretation if the χ^2 is significant. Besides, you will eventually want to write your own computer program to compute expected frequencies and the chi-square anyway!

10.3 Types of errors

We have seen how the researcher may make a decision about the alpha level to be reached on his data as a requirement for the rejection of the null hypothesis. Usually he presets $\alpha = .05$, or $\alpha = .01$, although sometimes in a pilot study based on a small sample he will preset $\alpha = .10$. In setting α before the data are analyzed, the researcher is exercising control over the probability of an incorrect rejection of the null hypothesis. If all the assumptions of the statistical model are met in the research, α will be the risk the researcher chooses to run of generating a spurious finding of the sort that *denies that chance alone could have produced an observed event which is in reality due to chance alone.* The researcher who performs many statistical tests at a preset $\alpha = .05$ is acknowledging his willingness to make an erroneous rejection of H_0 5 times out of 100. This error of rejecting the null hypothesis when it is in reality true is called a *Type I* error.

It is important to distinguish between a preset rejection level, α, and the probability of a computed test statistic. Alpha states a willingness to accept a certain probability for a Type I error. After the data are analyzed the probability for the test statistic states the actual probability of a Type I error in a specific research situation, if H_0 is rejected. For example, in the contingency analysis of sex versus response computed in the last section, the result was

$$\chi_1^2 = 16.66, \quad P > .999$$

Therefore, the obtained probability of H_0 is $p < .001$. Although the authors would have rejected H_0 for any $p \leq .05$, since they had preset $\alpha = .05$, the actual result indicates that for these data the risk of a Type I error is less than 1 chance in 1000. Alpha is a subjective fact, revealing the willingness to risk a kind of error in the inference game. Perhaps the best indicator of the risk of Type I errors actually run in a research program is the average of the actual probabilities for the test statistics for all rejected null hypotheses. In some fields of inquiry this prevailing risk would run close to .05, but in other fields it would be much less.

The goal of the game is to control the error rate in sequences of statistical inferences, and it is encouraging that statistics does provide even the novice with control over the risk of a Type I error. Unfortunately, there are too many examples in the educational research literature where the facade of this control is elegantly displayed, but the substance is missing. These are cases in which the data-collecting procedures are so far removed from correspondence with the assumptions of the statistical model that the act of asserting a probability for the null hypothesis is ridiculous. Many inference models are known to be "robust" with respect to some of their assumptions. That is, a model may be known to work pretty well with data that fail to meet one of the model's assumptions. Frequently this tolerable violation is a failure of normality for the sampled population, when the model assumes such normality. All statistical models deal with the behaviors of quantities computed on random variates, and *no model is robust when the element of randomness is missing from the data.* Far too often researchers collect data from available clusters of subjects and then test null hypotheses about

TYPES OF ERRORS 169

hypothetical populations on the basis of such data. Where there has been no randomization either in sampling or assigning to treatments, or where the stab at randomization has been obviously inappropriate, there can be no proper act of statistical inference. We hope you have ample realization that the other name of the statistical game is *inductive generalization*, and have long since intuited that the inductive leap from an observed regularity in a set of observations to a generalization about a total population of potential observations can produce reliable generalizations only when the sample is representative. Randomization is the key to achieving representativeness in samples.

If statistical inference about population-characterizing hypotheses is unwarranted when performed on nonrepresentative samples, why is the practice widespread? There are many excuses, usually involving budgetary considerations, but we all must be critical of this practice. It has loaded our literature with studies reporting unreliable generalizations, and has impeded the maturation of the behavioral sciences. As a consumer of research literature, you can carefully critique the sampling procedures of the articles you read, and more readily accept generalizations derived from data collected under adequate randomization conditions. If you become a researcher, you can strive more energetically than have many of your predecessors for adequate research designs with respect to this cardinal principle of randomization. If you should become a journal editor you would be in a position to make a monumental contribution to your science by insisting on adequate sampling procedures in the studies you publish.

Can test statistics and their associated probabilities play any proper role in researches based on available rather than random samples? Sometimes they can. Given the painful necessity of conducting a research program on an available sample, it is possible to employ statistical test probabilities as indices to the relative strengths of various trends in the data. Suppose we had arrived at the contingency table of the previous section in a study of questionnaire responses for 100 available subjects. The statement

$$\chi_1^2 = 16.66, \quad P > .999$$

alone is meaningless, because there is no population to which to generalize. But suppose that the example is the sex by response contingency table for the first item, and that the sex by response contingency table for the second item yielded the statement

$$\chi_1^2 = 5.42, \quad .95 < P < .99$$

Now a comparison of the two statistical test outcomes allows us to assert that the association between sex and response tendency *for these subjects* is stronger for the first item than for the second. In this usage, the chi-square and its probability are descriptive statistics, not inference statistics. Their role is in the organization and summarization of the data for a particular set of subjects, not in the testing of hypotheses about a population. Sometimes circumstances force the study of available samples for their own sake. Such inquiries should be described as *case studies* to distinguish them from sampling surveys and experiments. Rigorous application of this distinction would do much to clarify our literature.

If control of error rate is the goal of the game, why doesn't the researcher choose a very, very small value for α? It has been suggested that $\alpha = .01$ is gaining ascendancy over $\alpha = .05$ in the psychological literature. Why not shoot for universal acceptance of $\alpha = .001$, or some lesser alpha? The answer is that our enthusiasm for controlling the probability of a Type I error is tempered by our concern over the probability of another type of error. The *Type II* error is the failure to reject the null hypothesis when it is false.

Inference Errors

Type I: Reject a true H_0
Type II: Accept a false H_0

The preset alpha and the obtained probability for H_0 tell only about the probability of a Type I error. The probability of a Type II error is determined by the *power*, or *operating characteristics*, of each statistical inference model, and is a subject of excessive complexity for a first semester treatment. For the time being we want to point out to you only that the probability of a Type II error is inversely related to the preset probability of a Type I error. This means that as the researcher makes alpha smaller he increases his chances of failing to reject a false null hypothesis. In statistics, the probability of a Type II error is designated "beta" (β). So the rule is that decreasing α increases β. Increased control over one type of error involves decreased control over the other, for a given set of circumstances.

One way researchers try to resolve this dilemma is to change the circumstances. That is, the power of the test is related to other factors besides the α level chosen. We could, for example, decrease the probability of a Type II error by increasing the sample size, and still use the same α level. Also, as you gain more familiarity with a larger variety of statistical methods you will see that, although several different statistical tests might be appropriate for a given research hypothesis, some tests will be more powerful than others. In general, tests making more assumptions about the characteristics of the data (for example, that the variable is normally distributed) are less likely to result in a Type II error. When the data satisfy the assumptions of the more powerful statistical test, it should be used. Still another suggestion to reduce Type II errors is to state the more powerful one-tailed alternatives to the null hypothesis whenever possible. If that is not possible, you can report the

SUMMARY

Table 10.5 200 Empirical Sample Variances for 10 Random Fractions Generated by List 5.3 and Goodness of Fit to $N(M_{\sigma^2}, s_{\sigma^2}^2)$ ($M_{\sigma^2} = .073$, $s_{\sigma^2}^2 = .022$)

Class	Lower Limit	Upper Limit	Observed f	Expected f	$\dfrac{(O-E)^2}{E}$
1	.000	.036	7	10	.90
2	.037	.044	14	10	1.60
3	.045	.053	21	20	.05
4	.054	.060	22	20	.20
5	.061	.066	16	20	.80
6	.067	.072	22	20	.20
7	.073	.077	15	20	1.25
8	.078	.083	22	20	.20
9	.084	.090	23	20	.45
10	.091	.099	17	20	.45
11	.100	.107	8	10	.40
12	.108	.999	13	10	.90

$\chi_9^2 = 7.40$, $.10 < P < .90$

findings in terms of confidence limits and display the range of alternative hypotheses which are acceptable at the alpha level chosen.

As an example of a Type II error, consider a chi-square goodness of fit test for the hypothesis that sample variances for random samples of size 10 from a rectangular distribution distribute normally. We know that such sample variances do not arise from a normal distribution, yet Table 10.5 reveals that we cannot reject the null hypothesis of a normal source for 200 variances of 10 random fractions generated by List 5.3. The chi-square test simply isn't powerful enough for this job. This example casts a sobering light on our other goodness of fit tests, by the way.

10.4 Summary

For samples from a normal population, the null hypothesis

$$H_0: \hat{s}_x^2 - \sigma_x^2 = 0$$

can be tested by the chi-square statistic

$$\chi_{N-1}^2 = \frac{(N-1)\hat{s}_x^2}{\sigma_x^2}$$

A *contingency table* displays the association between two variables measured on the same group of subjects. In such a table, the cell in row i and column j contains the frequency of subjects who had the ith score on the first variable and the *j*th score on the second variable. The null hypothesis that there is no association (zero association) between the two variables in the population can be tested by a chi-square procedure. The total sample size, N, and the row totals (R_i) and column totals (C_j) are accepted as fixed. For a table with r rows and c columns there are $(r-1)$ times $(c-1)$ degrees of freedom. The expected frequency for a cell is

$$E_{ij} = R_i \cdot \frac{C_j}{N}$$

Then the test statistic is

$$\chi^2 = \sum_{i=1}^{r} \sum_{j=1}^{c} \left[\frac{(O_{ij} - E_{ij})^2}{E_{ij}} \right]$$

A *Type I error* is the rejection of the null hypothesis when it is true. A preset α represents a degree of *a priori* willingness to risk a Type I error. The actual $p(H_0)$ obtained from a statistical test represents the *a posteriori* probability of a Type I error, if H_0 is rejected.

A *Type II error* is the acceptance of the null hypothesis when it is false. The probability of such an error, β, decreases with increasing α, decreases with increasing N, and is less for one-tailed than for two-tailed tests on the same data.

EXERCISES AND QUESTIONS

1. The data of Appendix B represent a 0.5 percent sample of the corresponding grade and sex in the entire Project TALENT sample. Considering the latter a population, and given the fact that the variance, σ^2, for the abstract reasoning test (variable 15) for grade 12 males is 9.0, compute the variance estimate for your 0.5 percent sample of males and decide if the observed difference ($\sigma^2 - \hat{s}^2$) makes our sampling procedure suspect.

2. Using UNIVARIATE DISTRIBUTION, obtain the frequency distribution for variable 8 (Plan College Full Time) for both sexes separately. Using chi-square analysis, decide whether or not response to that item depends in part upon whether the respondent is a boy or a girl. That is, is there a significant relationship between sex and response to variable 8?

3. A very alert psychologist (some might say a troublemaker) recently published a study of a sample of statistical analyses that had appeared in several psychological journals. To qualify for the sample an analysis had to have led to a failure to reject the null hypothesis. For all these analyses the investigator computed the probability (β) of a Type II error, something the original authors had not done.

SUMMARY

Computing β for a statistical analysis is still a rather off-beat behavior for researchers, although we may hope that the opportunity to turn the chore over to a computer will lead to more popularity for it in the future. Anyway, our investigator found that the average β for the studies in his sample was approximately .50. This suggests that possibly in something like one-half of the cases where psychologists fail to find support for their research hypotheses in rejection of their null hypotheses the failure may be because of inadequate power in the statistical analyses. What do you think the implications of this finding might be for researchers and for editors of research journals? Perhaps one member of the class should read this article and report on it (Cohen, 1962).

CHAPTER ELEVEN

Small Sample Statistics

11.1 The t distribution

We have seen that for some statistics the sampling distribution is normal. For other statistics, we have found the binomial distribution or the chi-square distribution to be relevant. As you learned in Chapter 8, the sampling distribution of means based on samples drawn from the same population is generally normal. The exception is when we are dealing with very small samples. This chapter considers the problem of making inferences about means when the samples are small.

There are actually two approaches to the problem of making inferences about means based on small samples. The Pearsonian approach uses what is called a critical ratio which is normally distributed and which is defined by the following formula.

$$z = \frac{M_x - \mu}{\sigma_M}$$

The Pearsonian approach assumes a knowledge of the population standard deviation from which the population's standard error of the mean is computed. The problem is that we seldom know the population standard deviation. For that reason, another procedure called Fisherian or "Studentized" statistics has been developed which depends only on the variance estimate computed from the sample in hand. The Fisherian approach utilizes the unbiased estimate of the standard deviation, and always assumes X is normally distributed. Let us define the statistic t as the difference between the sample mean and the population mean, divided by the unbiased estimate of the standard error of the mean. The formula would be:

$$t = \frac{M_x - \mu}{s_M} \qquad (11.1.1)$$

The sampling variance of the mean is defined as follows:

$$s_M{}^2 = \frac{\sum (X - M_x)^2}{N(N-1)} = \frac{\hat{s}_x{}^2}{N}$$

Note that the only difference between this t formula and the formula for the

THE t DISTRIBUTION

critical ratio is that in this case an unbiased estimate of variance is substituted for the known population variance in the critical ratio. The above definitional formula for *t* might be used by a teacher who wanted to know whether or not her class of students differed from the school population on reading comprehension ability. If the school mean were 100, and the 20 students in her class had a mean of 105.2, she could compare the school population mean with her class mean to see whether or not her 20 students could be considered a random sample from the entire school. If she didn't have the school

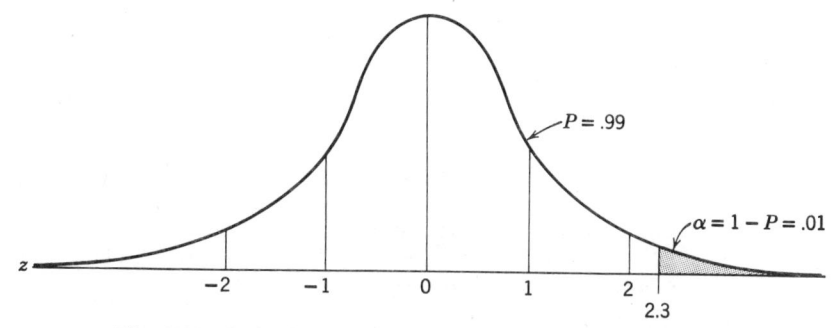

Fig. 11.1 Probability of obtaining a value of z greater than 2.3.

variance, she would estimate it from her sample of 20 students. The appropriate estimate would be the unbiased estimate. If her variance estimate turned out to be 100, this would mean her $t = (105.2 - 100)/\sqrt{100/20} = 5.2/2.24 = 2.3$. The teacher has now computed a statistic called *t* and, if she knew the sampling distribution of this statistic, she could decide whether this was an unusually large *t*.

For the moment, let us assume that *t* is distributed as a standard normal deviate. What is the probability of getting a *t* as large as or larger than the obtained value? Looking at our distribution of z in Table 7.1 we find that the probability of obtaining a z this large or larger is about .01. The areas involved with this z are illustrated in Figure 11.1. Now let us compare this result to the table values for *t* found in Table 11.1. Inspection of Table 11.1 indicates that there is not one but many *t* distributions. The column n.d.f. defines the number of degrees of freedom for a particular *t* distribution. For the moment we don't know where these *t* distributions came from, but let us assume that they are the sampling distributions for formula 11.1.1, where the particular distribution depends upon sample size (n.d.f. = $N - 1$ in this case). Entering this table with 19 degrees of freedom we determine the areas or proportions for the curve indicated in Figure 11.2. Comparing Figures 11.1 and 11.2 we get our first sign that the *t* distribution is very similar to but not identical with the normal distribution. Notice that there are

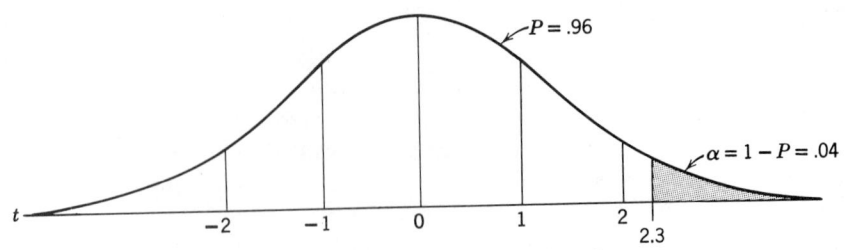

Fig. 11.2 Probability of obtaining a value of t greater than 2.3 with n.d.f. = 19.

Table 11.1 Cumulative Probabilities for t^a

n.d.f. P:	.001	.01	.05	.10	.90	.95	.975	.99	.999
1	−318.	−31.8	−6.31	−3.08	3.08	6.31	12.7	31.8	318.
2	−22.3	−6.97	−2.92	−1.89	1.89	2.92	4.30	6.97	22.3
3	−10.2	−4.54	−2.35	−1.64	1.64	2.35	3.18	4.54	10.2
4	−7.17	−3.75	−2.13	−1.53	1.53	2.13	2.78	3.75	7.17
5	−5.89	−3.37	−2.02	−1.48	1.48	2.02	2.57	3.37	5.89
6	−5.21	−3.14	−1.94	−1.44	1.44	1.94	2.45	3.14	5.21
7	−4.79	−3.00	−1.90	−1.42	1.42	1.90	2.37	3.00	4.79
8	−4.50	−2.90	−1.86	−1.40	1.40	1.86	2.31	2.90	4.50
9	−4.30	−2.82	−1.83	−1.38	1.38	1.83	2.26	2.82	4.30
10	−4.14	−2.76	−1.81	−1.37	1.37	1.81	2.23	2.76	4.14
11	−4.03	−2.72	−1.80	−1.36	1.36	1.80	2.20	2.72	4.03
12	−3.93	−2.68	−1.78	−1.36	1.36	1.78	2.18	2.68	3.93
13	−3.85	−2.65	−1.77	−1.35	1.35	1.77	2.16	2.65	3.85
14	−3.79	−2.62	−1.76	−1.35	1.35	1.76	2.15	2.62	3.79
15	−3.73	−2.60	−1.75	−1.34	1.34	1.75	2.13	2.60	3.73
16	−3.69	−2.58	−1.75	−1.34	1.34	1.75	2.12	2.58	3.69
17	−3.65	−2.57	−1.74	−1.33	1.33	1.74	2.11	2.57	3.65
18	−3.61	−2.55	−1.73	−1.33	1.33	1.73	2.10	2.55	3.61
19	−3.58	−2.54	−1.73	−1.33	1.33	1.73	2.09	2.54	3.58
20	−3.55	−2.53	−1.73	−1.33	1.33	1.73	2.09	2.53	3.55
22	−3.51	−2.51	−1.72	−1.32	1.32	1.72	2.07	2.51	3.51
24	−3.47	−2.50	−1.71	−1.32	1.32	1.71	2.06	2.50	3.47
26	−3.44	−2.48	−1.71	−1.32	1.32	1.71	2.06	2.48	3.44
28	−3.41	−2.47	−1.70	−1.31	1.31	1.70	2.05	2.47	3.41
30	−3.39	−2.46	−1.70	−1.31	1.31	1.70	2.04	2.46	3.39
60	−3.23	−2.39	−1.67	−1.30	1.30	1.67	2.00	2.39	3.23
120	−3.16	−2.36	−1.66	−1.29	1.29	1.66	1.98	2.36	3.16
∞	−3.09	−2.33	−1.65	−1.28	1.28	1.65	1.96	2.33	3.09

[a] Values selected from Abramowitz and Stegun (1964), p. 990.

COMPARISON OF TWO MEANS

slightly more cases falling above the obtained t value than is the case if we consider a z of the same size. This suggests that the distribution of t is slightly platykurtic (flatter). The extent to which the distribution of t is platykurtic depends on the sample size. The smaller the sample, the flatter the curve. As the sample sizes get larger and larger, the t distribution gets closer and closer to the normal distribution. This can be seen by comparing the values of t for an infinite number of degrees of freedom with the normal curve table and noticing that they are identical.

11.2 Confidence intervals

If we rewrite our definitional formula 11.1.1 as $\mu = M_x \pm ts_M$ we then have a way of talking about the confidence we have in estimating a population mean from the obtained sample mean, given the obtained unbiased estimate of the population variance. Choosing a value of t equal to 2.1,[1] because this is the absolute value of t with 19 degrees of freedom which is exceeded only 5 percent of the time, we can obtain the confidence limits for the class mean which the teacher obtained above. (Note that we obtained the $t_{19} = 2.1$ for $\alpha = .05$ from the $P = .975$ column of Table 11.1, because we want a two-tailed α level, that is, we want $p\,|t_{19}| \leq .05$.) In that example the sample mean was 105.2 and the standard error of the mean was 2.24. Multiplying S_M by 2.1 we find an interval of 4.7. Thus, the lower bound of the interval is the mean minus 4.7 or 100.5 and the upper limit of the interval is the mean plus 4.7 or 109.9. Now the question arises as to what we can say about this confidence interval. The value of t that we chose was at the two-tailed .05 level. Therefore, we can say that the mean of the population from which this sample was drawn probably lies somewhere in between these two interval limits, 100.5 and 109.9. Or, saying it more precisely, 95 percent of the time that we estimate confidence intervals using t equal to 2.1, the population mean will lie within the established confidence interval.

Looking back at our formula, we see that the larger the t we select the larger the interval will be. Also, as the variance increases, the standard error of the mean increases, and the interval becomes larger. The standard error of the mean, in turn, is dependent upon sample size. Thus the larger the sample, the more precisely you can define the limits within which the population mean probably lies.

11.3 Comparison of two means

As you become more familiar with educational research, you will realize that most of the interesting and important research that is done generally involves contrasts, or the making of comparisons. It is generally not of much interest to know only that a particular sample had a mean of 50 on test X.

[1] Rounded from 2.09.

One reason that this is often not of interest is that our scales are generally not ratio scales and so the absolute value has very little meaning. When this mean of 50 is placed on a norming scale and we find that the class is performing above its age group, then the results become a little more interesting. In this way we have made a comparison between the sample in hand and other available estimates of population parameters. We might want to ask, however, whether this sample could be considered a random sample from the population for which the population mean is available or whether this sample came from some other population with a different mean. To do this we would use formula 11.1.1 above.

A more frequent case is where we have two sample means and the question is, "Can these two samples be considered as random samples from the same population?" Here we are contrasting two means and are wondering whether or not we can talk about the group having the higher mean as being better than the group having the lower mean. Of course we can look at the two means and decide which is the larger number. However, that is not our research question. If these two groups represent two different treatment groups, we want to know whether those students receiving treatment A do better than the students receiving treatment B. If we randomly assign the students to the two treatment groups, and then give them a test without treatment, scores will be different just due to sampling variation. What we want to know is how much sampling variation should be expected and when does this difference become large enough for us to be willing to say that it is due to treatment differences rather than just differences due to sampling error.

The formula we introduced before for the t distribution does not provide for the comparison of two sample means. Fortunately, however, an Englishman who wrote under the pseudonym of "Student" has provided us with such a procedure. He proposed in 1908 the following formula for testing the difference between two means:

$$t = \frac{(M_1 - M_2)}{S_{M_1 - M_2}}, \quad \text{n.d.f.} = N_1 + N_2 - 2 \quad (11.3.1)$$

Two assumptions Student's formula depends upon are (1) that the populations are normal and (2) that the populations have the same variance, so that $\sigma_1^2 = \sigma_2^2 = \sigma^2$.

We want to ask whether the difference between two observed means is significantly different from zero. Again we are determining whether the difference is unusually large by comparing the observed difference to the standard error of the differences. This latter, of course, is the standard deviation for the distribution of differences between two sample means when the samples are drawn from a common population. This standard error of the difference between two means should be computed from the following formula:

$$S_{M_1-M_2} = \sqrt{\left(\frac{\sum x_1^2 + \sum x_2^2}{N_1 + N_2 - 2}\right)\left(\frac{1}{N_1} + \frac{1}{N_2}\right)} \quad (11.3.2)$$

Notice that this formula pools the accumulations (sums of squares) from the two samples, and is different from formula 8.3.1 for a standard error of a mean difference we used in Chapter 8. This new formula is justified by the assumption that $\sigma_1^2 = \sigma_2^2$, which we did not make in Chapter 8.

So far in this chapter we have asked you to take a great deal on faith. It is now time we do some demonstrations to show you that the sampling distribution of formulas 11.1.1 and 11.3.1 above do in fact follow the well-known distribution t as indicated in Table 11.1. Once again we will turn to the computer and the Monte Carlo method to examine the nature of this distribution.

11.4 Monte Carlo demonstrations of t distribution

List 11.1 is a program for computing the statistic t based upon the difference between two sample means. As presented there, the program is set to draw 200 pairs of random samples of five subjects from each of two normal populations. The program parameter XMD determines the extent to which the two population means differ. When XMD is set equal to 1.00, for example, one population mean is 1.00 and the other is zero. The standard deviation within both populations is always set at 1.00.

Table 11.2 summarizes the outcomes of a computer run using a mean

Table 11.2 Empirical t_8 with XMD $= 0.0$ from List 11.1 and Goodness of Fit to Theoretical t_8

Class	Lower Limit	Upper Limit	Observed f	Expected f	$\frac{(O-E)^2}{E}$
1	−9.000	−1.860	10	10	.000
2	−1.859	−1.400	8	10	.400
3	−1.399	−.889	23	20	.450
4	−.888	−.546	17	20	.450
5	−.545	−.262	14	20	1.800
6	−.261	−.000	26	20	1.800
7	.000	.261	23	20	.450
8	.262	.545	20	20	.000
9	.546	.888	19	20	.050
10	.889	1.399	24	20	.800
11	1.400	1.859	6	10	1.600
12	1.860	9.000	10	10	.000

$\chi_{11}^2 = 7.80$, $.10 < P < .90$

difference (XMD) of zero. Therefore, this empirical distribution of t describes the sort of outcomes we would expect to get in repeated experiments where the two samples differed only by chance. Notice that the 200 observed t's follow rather close to the t's expected on the basis of the tabled theoretical distributions. For example, Table 11.1 tells us that with 8 degrees of freedom we expect to find 10 out of 200 (5 percent) of them to be 1.86 or larger. This is exactly what happened in this Monte Carlo experiment. Thus we have a clear demonstration of a characteristic of the t-test which many users of statistical inference do not appear to comprehend until too late. That is, the fact that there is nothing magical about the t value which is associated with the .05 level—where the cumulative probability is .95. In this application .05 is simply the proportion of times that t's based on samples drawn from two populations with the same mean (or from the same population) will produce a t value this large or larger. That is, 5 percent of the time that we apply formula 11.1.1 to two samples with only random differences, the obtained t will exceed the tabled value of t where $P = .95$.

Table 11.3 Empirical t_8 with XMD = 1.0 from List 11.1 and Goodness of Fit to Theoretical t_8

Class	Lower Limit	Upper Limit	Observed f	Expected f	$(O - E)^2 / E$
1	−9.000	−1.860	82	10	518.40
2	−1.859	−1.400	38	10	78.40
3	−1.399	−.889	32	20	7.20
4	−.888	−.546	19	20	.05
5	−.545	−.262	12	20	3.20
6	−.261	−.000	9	20	6.05
7	.000	.261	4	20	12.80
8	.262	.545	3	20	14.45
9	.546	.888	0	20	20.00
10	.889	1.399	1	20	18.05
11	1.400	1.859	0	10	10.00
12	1.860	9.000	0	10	10.00

$\chi_{11}^2 = 698.6$, $.999 < P$

In Table 11.3 are the Monte Carlo results of applying the t-test to 200 sample pairs drawn from two populations with 1 standard deviation difference in means (that is, XMD = 1.00). Notice here that only 41 percent of the time (82 out of 200) was the obtained t value large enough to be significant at the .05 level of significance. This illustrates the low power of the t test when very small samples are used. The 59 percent of the time that the t value did not reach the .05 level is the proportion of times a Type II error occurred in this

11.5 Application of *t*-test

In the grade 12 male data of Appendix B there are 12 boys who are 16 years old and 8 who are 19 years old. Thus we have two very small samples, one for each of these two age groups, Therefore, if we wanted to hypothesize that for a given grade the younger students will do better on a reading comprehension test than the older students, the *t* distribution would provide the appropriate basis for comparing these two sample means.

List 11.2 provides a program for computing the means and standard deviations for two groups together with the resulting *t* and the number of degrees of freedom associated with that *t*. The printout to that program is summarized in Table 11.4. There we see that the 16-year-olds did indeed do

Table 11.4 Comparison of Reading Comprehension for Sixteen- and Nineteen-Year-Old Twelfth Graders

	Sixteen-Year-Olds	Nineteen-Year-Olds
N	12	8
Means	35.08	25.25
Standard Deviation	7.59	7.69
	$t_{18} = 2.83$	
	$P(t_{18}) > .990$	

better on the reading comprehension test than did the 19-year-olds in the twelfth grade. Entering Table 11.1 with 18 degrees of freedom, we find that the obtained *t* of 2.83 has a *P* between .990 and .999. Therefore, we can be quite confident that these two samples come from populations with different reading comprehension means. There is less than 1 chance in a 100 that the superiority of the 16-year-olds over the 19-year-olds could be attributed to sampling error.

The research hypothesis that 16-year-olds would do better on reading comprehension than 19-year-olds in the same grade is based on the assumptions that 16-year-olds would tend to be those boys who were moved ahead of their age group in school because of their superior scholastic aptitude. Also, the 19-year-olds would tend to be those who were held back at some point in their schooling because of scholastic difficulties. Turning to Appendix B variables, can you suggest some variables where the 19-year-olds would have a higher mean than the 16-year-olds? If you can develop a reasonable basis for a directional hypothesis of this type, sort out the 16- and 19-year-olds from your male deck, and using List 11.2 see if your hypothesis holds up. Also, would you expect different age trends for girls than for boys?

11.6 Summary

The t distribution gives us a basis for inference tests on means of very small samples, for which the sampling distributions of means and mean differences are not quite normal.

A confidence interval can be established around a sample mean by using

$$M_x \pm t_n(\alpha) S_M$$

where $t_n(\alpha)$ is the value of t with n degrees of freedom for which $p\,|t| \leq \alpha$, and S_M is the standard error of the mean. Then α is the probability that the interval does not include the population mean, and $1 - \alpha$ is the confidence we have that the interval does include μ.

The most useful application of t is in the testing of the significance of an observed difference between the means of two independent random samples. The null hypothesis is that the two populations sampled have the same mean. The enabling assumptions are (1) normal populations, with (2) equal variances.

The Monte Carlo study verifies the suitability of the t distribution for testing this null hypothesis. It also demonstrates the proportion of Type II errors for a given actual difference between the two population means.

EXERCISES AND QUESTIONS

1. For 20 degrees of freedom, what proportion of the t distribution lies between t values of -1.33 and $+1.33$?

2. An experiment was conducted in which 16 students taught by method A had a mean of 65 on the test while 16 students with method B had a mean of 45. If the standard error of the difference between these two means is 10, what can be said about the relative effectiveness of methods A and B, assuming test X was an appropriate test in this experimental situation?

3. What more would you like to know about the experiment of Question 2 in order to properly evaluate it?

4. We have noted that Equation 11.3.2 is an estimate of the standard error of the difference in means that is based on the assumption of a common population

SUMMARY

variance, $\sigma_1^2 = \sigma_2^2 = \sigma^2$. Therefore, in general $S_{M_1-M_2}$ as defined in (11.3.2) is not algebraically equivalent to the definition given in (8.3.1), namely

$$S_{M_1-M_2} = \sqrt{S_{M_1}^2 + S_{M_2}^2}$$

which is for the assumption $\sigma_1^2 \neq \sigma_2^2$. However, when $N_1 = N_2 = N$ formula 11.3.2 is equivalent to formula 8.3.1, so that the pooled samples variance error is the same as the sum of the individual samples variance errors. This is important because it suggests that when the researcher is anxious about assuming $\sigma_1^2 = \sigma_2^2$ he should employ equal sample sizes. Can you do the algebra to show that (11.3.2) is equivalent to (8.3.1) if $N_1 = N_2 = N$?

5. Try to phrase in your own words the meaning and the justification of the statement that $\beta = .59$ in the authors' run of the Monte Carlo experiment on t with 8 degrees of freedom 2nd XMD = 1.0. Also, phrase a statement about the obtained powers of t_8 in your runs of this experiment with various values of the difference in means, XMD.

6. Notice that our table of the t distribution (Table 11.1) gives values of t corresponding to selected values of P, the cumulative probabilities. Our chi-square table (Table 9.1) also reports for selected values of P. When interpreting a test statistic, whether it is a χ^2 or a t, you want values of α, not of P. What is the relationship of α to P, and how can you get α from P?

```
C       LIST 11.1    MONTE CARLO ON T TEST
C
C       THIS PROGRAM COMPUTES T WITH 8 DEGREES OF FREEDOM FOR 200 RANDOM
C       SAMPLES OF 10 SUBJECTS EACH, 5 FROM EACH OF TWO NORMAL POPULATIONS
C       WITH THE SAME VARIANCE (UNITY) AND MEANS SEPARATED BY PARAMETER
C       XMD. WHEN THE PROGRAM IS RUN WITH XMD EQUAL TO ZERO, THE NULL
C       HYPOTHESIS IS TRUE AND THE EMPIRICAL DISTRIBUTION OF T SHOULD BE
C       A RANDOM SAMPLE FROM THE THEORETICAL DISTRIBUTION OF T. THE POWER
C       OF T IS DEMONSTRATED FOR 6 VALUES OF XMD.
C       THE SINGLE DATA CARD GIVES AN 8-DIGIT STARTING NUMBER FOR RANDOM
C       IN COLUMNS 1-8.
C       REQUIRED SUBROUTINES ARE RANDOM, RANDEV, UNIDIS, AND GOOFIT.
C
        DIMENSION    T(1000),   F(100),   IC(100),   IE(100)
C
        WRITE (6,4)
 4      FORMAT (53H1 MONTE CARLO ON T WITH 8 DEGREES OF FREEDOM             )
        N = 200
        Y = RANDOM(1)
C
        F(1) = -9.000
        F(2) = -1.859
        F(3) = -1.399
        F(4) = -.888
        F(5) = -.545
        F(6) = -.261
        F(7) = .000
        F(8) = .262
        F(9) = .546
        F(10) = .889
        F(11) = 1.400
        F(12) = 1.860
        M = 12
        IE(1) = 10
        IE(2) = IE(1)
        IE(11) = IE(1)
        IE(12) = IE(1)
        IE(3) = 20
        IE(4) = IE(3)
        IE(5) = IE(3)
        IE(6) = IE(3)
        IE(7) = IE(3)
        IE(8) = IE(3)
        IE(9) = IE(3)
        IE(10) = IE(3)

        XMD = .00
        DO 2     L = 1, 6
        DO 3     K = 1, N
        SUM1 = 0.0
        SSQ1 = 0.0
        SUM2 = 0.0
        SSQ2 = 0.0
C
        DO 5     J = 1, 5
        Z = RANDEV(0)+ XMD
        SUM1 = SUM1 + Z
 5      SSQ1 = SSQ1 + Z * Z
        DO 6     J = 1, 5
        Z = RANDEV(0)
        SUM2 = SUM2 + Z
 6      SSQ2 = SSQ2 + Z * Z
C
        SSQ1 = SSQ1 - SUM1 * SUM1 / 5.0
        SSQ2 = SSQ2 - SUM2 * SUM2 / 5.0
        SXMD = (SUM1 - SUM2) / 5.0
        SED = SQRT (((SSQ1 + SSQ2) / 8.0) * .40)
 3      T(K) = SXMD / SED
C
        WRITE (6,7)     XMD
 7      FORMAT (18HOMEAN DIFFERENCE = F10.5)
        CALL UNIDIS (N, M, T, F, IC)
        CALL GOOFIT (M, F, IE, IC)
 2      XMD = XMD + .20
C
        CALL EXIT
        END
```

SUMMARY

```
C      LIST 11.2   T TEST
C           THIS PROGRAM COMPUTES STUDENTS T FOR TWO GROUPS ON ONE
C      VARIABLE. THE FIRST CONTROL CARD READ BY PROGRAM CONTAINS IN COLS.
C           1 - 3 = N1,  NUMBER IN GROUP 1
C           4 - 6 = N2,  NUMBER IN GROUP 2
C      PROGRAM EXPECTS VARIABLE X TO BE IN COLUMNS 11 - 12.
       RIT 7, 2, N1, N2
2      FORMAT (2I3)
       EN1 = N1
       EN2 = N2
       DF = N1 + N2 - 2
       SUM1 = 0.0
       SSQ1 = 0.0
       SUM2 = 0.0
       SSQ2 = 0.0
       DO 4 J = 1, N1
       RIT 7,3,X
3      FORMAT(10X F2.0)
       SUM1 = SUM1 + X
4      SSQ1 = SSQ1 + ( X * X )
       SSQ1 = SSQ1 - (SUM1 * SUM1 / EN1 )
       DO 5 J = 1, N2
       RIT 7,3,X
       SUM2 = SUM2 + X
5      SSQ2 = SSQ2 + ( X * X )
       SSQ2 = SSQ2 - ( SUM2 * SUM2 / EN2 )
       XM1 = SUM1 / EN1
       XM2 = SUM2 / EN2
       SXMD = XM1 - XM2
       SED = SQRTF(((SSQ1 + SSQ2) / DF) * ( 1.0/EN1 + 1.0/EN2))
       T = SXMD / SED
       SD1 = SQRTF(SSQ1/(EN1 - 1.00))
       SD2 = SQRTF(SSQ2/(EN2 - 1.00))
       WOT 6,6
6      FORMAT(1H1,11X30HGROUP ONE           GROUP TWO )
       WOT 6,7,N1,N2,XM1,XM2
7      FORMAT(4HON = 14XI3,18XI3/ 8HOMEANS = 6XF7.2, 14XF7.2)
       WOT 6,8,SD1,SD2
8      FORMAT(12HOSTD. DEV. =  F9.2, 12XF9.2)
       WOT 6,9,T,DF
9      FORMAT(13HOSTUDENTS T = F6.2, 11H WITH NDF = F5.0)
       END
```

CHAPTER TWELVE

Exact Randomization Tests

12.1 The randomization outcomes of an experiment

In *The Design of Experiments*, R. A. Fisher discusses the possibility of computing all the outcomes of a small-scale experiment which could have occurred by different random assignments of subjects to treatments, in order to assess the probability of the actual outcome observed (1950, p. 43). That is, Fisher suggested that each subject's criterion score be thought of as an intrinsic possession of that subject which is not modified by membership in a particular treatment group. Assuming that the dependent variable scores for the sample of subjects cannot be affected by any of the treatments, every possible arrangement of subjects in treatment groups is made and the test statistic is computed for each arrangement. What results is a complete distribution of the values of the test statistic that could occur if there were no treatment effect and if the outcome of the experiment were due strictly to the chance arrangement of the subjects in the treatment groups produced by random assignment. Then the probability of a test statistic equal to or greater than the one that has been computed on the actual observed outcome of the experiment can be determined from the place of the actual statistic in this distribution of exact randomization outcomes.

This chapter is devoted to an exploration of Fisher's notion not because of its practical importance (although it has some), but because it thoroughly dramatizes the essential role of random assignment in an experiment. One eminent statistician has observed that: "Tests of significance in the randomized experiment have frequently been presented by way of normal law theory whereas their validity stems from randomization theory" (Kempthorne, 1955). We have emphasized the central-limit theorem and the normal distribution in this text, and properly so, but if there is one single principle we want to have you retain from this course above all others it is the principle that *randomization is the absolutely indispensable element of statistical models for research*. To reinforce this principle, we consider here how a statistical model for inference about an experimental outcome can be built on the sole assumption of random assignment.

As you continue to explore statistics you will become aware of one of its divisions which is called "nonparametric statistics." Since Siegel's excellent text appeared in 1956 there has been a spread of interest among behavioral scientists in inference models that do not require special assumptions about the forms of the population distributions that are sampled. Frequently data have to be collected in ways that simply do not warrant the assumption of interval scales and normal distributions, and it is well that models that avoid these assumptions are available. Our point here is that exact randomization tests are the ultimate nonparametric designs, since they assume nothing whatever about the parent distributions from which the observations are samples. They only assume that observations are independent. Siegel puts it this way: "With a randomization test, we can obtain the exact probability under H_0 associated with the occurrence of our observed data, and ... without making any assumptions about normality or homogeneity of variance" (1956, p. 88).

Table 12.1

	War College	Princeton
Admirals	A B	C D
Scores	5 3	4 2
Means	$M_1 = 4$	$M_2 = 3$

$$t_2 = \frac{M_1 - M_1}{\sqrt{\left(\frac{\sum x_1^2 + \sum x_2^2}{N_1 + N_2 - 2}\right)\left(\frac{1}{N_1} + \frac{1}{N_2}\right)}} = \frac{4 - 3}{\sqrt{\left(\frac{2+2}{2+2-2}\right)\left(\frac{1}{2}+\frac{1}{2}\right)}} = .71$$

To see the trick involved in an exact randomization test, it is helpful to play with a nearly trivial example. Suppose the Pentagon makes a random choice of two out of four available admirals, and sends the chosen pair (A and B) to the War College to take a six months course in geopolitics. The other pair (C and D) is sent to Princeton for a semester of tutorial study of the same topic. Time passes, the courses of study are completed, and the four admirals take a test of their understanding of geopolitics. The test scores for the four admirals, and the t test for the school effect, are shown in Table 12.1. Entering Table 11.1 with 2 degrees of freedom we find that a $|t| = .71$ is not unusually large ($.10 < P < .90$) so that the null hypothesis of zero school effects cannot be rejected. Suppose the obtained test scores represent the number of strategic blunders committed in a situational test which permits a maximum of five such blunders. There is reason to suspect some sort of a skewed multinomial distribution in the population of admirals, but certainly not a normal distribution. Anyway, the sample is not a random sample from the population, and the only legitimate research question concerns the extent to which chance alone explains the outcomes of this experimental case study. That is, with proper analysis we may expect to generalize about the school effect for these subjects. We will not be able to generalize to a population of subjects.

We have indicated what a t table says about the probability of the null hypothesis of zero school effect on the basis of these data. It says that there is no evidence against the null hypothesis. Unfortunately, it is presumptuous to use the theoretical distribution of t for comparing the computed t because the violations of the assumptions of the model for statistical inference are so extreme. There has, however, been random assignment of available subjects to experimental treatments, and it is a legitimate exercise in statistical inference to determine the two-tailed probability of a $|t|$ equal to or greater than the observed $|t|$ in the complete distribution of possible t outcomes of the randomization, assuming no treatment effect.

How many possible t values could have resulted from different randomizations of the subjects? The question is best seen as that of how many different combinations are possible of four things taken two at a time. This is because the Pentagon had four admirals available, and assigned two of them at random to the War College and the other two to Princeton. You will recall from Chapter 6 that combinations of four things taken two at a time, $\binom{4}{2} = \frac{4!}{2!\,2!}$, which works out to be 6. The six ways of assigning two of four admirals to the War College, which are the six possible outcomes of a random draw of two elements from a set of four elements, are shown in the following table, along with the accompanying Princeton pair for each case.

War College	Princeton
A B	C D
A C	B D
A D	B C
B C	A D
B D	A C
C D	A B

The exact randomization test requires the computation of a t value for each possible assignment, assuming that the admirals' test scores are fixed possessions of the admirals themselves and are not modified by the treatments. That is, the six possible t values are computed on the assumption that the test score for Admiral A is 5, regardless of the group to which he is assigned. Likewise the test score for B is fixed as 3, for C as 4, and for D as 2.

In Table 12.2 the six possible randomization outcomes are listed in order of the computed t values, from high to low. Since each of the six possible outcomes is equally likely to occur, the probability of any one t value is 1/6 (that is, $p(t) = .167$). The final column gives the cumulative probabilities for the possible t values. This column is similar to Table 11.1. Note, however, that these values of $P(t)$ in Table 12.1 are *not* based on the theoretical distribution of t (as in Table 11.1), but are based only on the empirical distribution of

Table 12.2 Randomization Outcomes for Admirals Experiment

Possible Randomization Outcomes	t	$p(t)$	$P(t)$
AC:BD	2.83	.167	1.000
AB:CD	.71	.167	.833
AD:BC	.00	.167	.667
BC:AD	.00	.167	.500
CD:AB	−.71	.167	.333
BD:AC	−2.83	.167	.167

the possible randomization outcomes for t given this particular set of data. This distribution makes it apparent that the probability of a t equal to or greater than our outcome is $2(.333) = .666$ so there is insufficient evidence in the data against the null hypothesis that $\mu_1 - \mu_2 = 0$. Another way of seeing this is to observe that there are four ways out of six possible outcomes to get the result equal to or greater than a $|t|$ of .71, when H_0 is true and the observed $|t|$ is due to chance alone. Note that reference to the Student t table produced approximately the same result. This seems to be a tribute to the robustness of the t-test. We are not arguing against the use of the standard t-test in situations which depart from its assumptions. We are merely trying to emphasize the role of randomization in every statistical design for experiments.

The complete distribution of exact randomization outcomes for the hypothetical experiment was easily computed because it contains only six points. Consideration of the number of values in the distributions for experiments of more practical sizes raises a serious problem, however. The general rule for two-treatment uncorrelated-observations designs is that the number of ways to assign half of a total sample of size N to one of the treatments is

$$\binom{N}{N/2} = \frac{N!}{[(N/2)!]^2} \qquad (12.1.1)$$

The implications for experiments involving 10, 20, 50, and 100 subjects are:

$$\binom{10}{5} = \frac{10!}{(5!)^2} = 252$$

$$\binom{20}{10} = \frac{20!}{(10!)^2} = 1.8476 \times 10^7$$

$$\binom{50}{25} = \frac{50!}{(25!)^2} = 1.6371 \times 10^{14}$$

$$\binom{100}{50} = \frac{100!}{(50!)^2} = 1.0089 \times 10^{29}$$

For ordinary sized samples the distribution of randomization outcomes is too extensive to bear computing! An attractive strategy is to compute a random sample of the possible randomization outcomes, say of size 200, and refer the observed outcome of the real experiment to this estimate of the randomization outcomes distribution. When we follow this strategy we no longer have an *exact* randomization test, because we do not get the precise probability of the observed t in the population of possible values for t, but rather get an approximation to the probability for the exact test.

Given the dependent-variable scores obtained by N subjects (assuming $N \geq 10$) in a two-treatment experiment, how can we construct a random sample of size 200 of the large number of possible randomizations of the subjects? If we can program a procedure for creating one of the possible combinations of N subjects taken $N/2$ at a time, at random, we shall be able to operate the procedure 200 times. Perhaps you recall that in Chapter 5 a method was described for computing random digits in the range 1 through N. A single random digit in this range is computed by calling RANDOM(J) for a random fraction, multiplying the fraction by N, "fixing" the product, and adding 1 to the resulting digit. In FORTRAN this operation is:

IX = RANDOM(0) * EN + 1.0

A list of $N/2$ such random digits would assign a random half of the subjects to the first treatment group, providing all the digits were different. Operating the procedure $N/2$ times may produce a list with some repetitions in it, which is not permissible. It is necessary to check each digit to see if it has already been added to the list, in which case it is ignored, and to continue generating digits until there are $N/2$ different digits in the list. These are then taken as the subscripts identifying the one-half of the subjects to be placed in the first group. The remaining $N/2$ subjects are treated as the second group.

THE RANDOMIZATION OUTCOMES OF AN EXPERIMENT

A t is computed for this randomization of the data. Please study List 12.1 carefully for the details of the operation. Note that a vector, $NV(J)$, of length N is zeroed initially. Then each digit generated is treated as a subscript for locating an element of $NV(J)$. Letting K be the random digit, a test is made to determine whether $NV(K) = 0$. If $NV(K) = 0$, the implication is that the random digit is not repetitious, and it is stored in the element of $NV(J)$ it has identified, by the replacement $NV(K) = K$. If $NV(K) \neq 0$, the digit K is ignored. A count is kept of the number of digits accepted, until $N/2$ have been accepted. When $N/2$ random digits have been accepted, the subscripts of the elements of $NV(J)$ which are nonzero identify the subjects to be assigned to the first group, and the subscripts of the elements of $NV(J)$ which are still zero in contents identify the subjects to be assigned to the second group.

The operation is repeated 200 times, yielding 200 t values, which the program orders from smallest to largest. The *actually obtained t*, which the program also reports, is compared by the user with the ordered randomization outcomes. A nice thing about using 200 outcomes is that .01 times the order number (or rank) of the randomization outcome equal to or closest to (on the small side) the absolute value of the obtained t is the two-tailed probability of the actual outcome of the experiment on the null hypothesis that randomization alone explains the observed group difference. Why is this so? How would you obtain the one-tailed t probability if you had the *a priori* hypothesis that $\mu_1 > \mu_2$ and the t for actual outcome was positive? What if the t for actual outcome was negative?

The authors borrow an example from a frequently used reference book on applied statistics (Wert, Neidt, and Ahmann, 1954, p. 131). In this example, the data for which appear in Table 12.3, a group of 20 students was drawn randomly from a pool of 40 available students, to constitute an experimental treatment group. The experimental group received "supplementary mimeographed exercises designed to be helpful with respect to the application of principles to new situations." The 20 subjects in the control group received the same instructional treatment with the exception that they did not receive the supplementary mimeographed exercises. When a test measuring application of principles was administered to the two groups after the differential treatments, the mean for the experimental group was 27.9 and the mean for the control group was 22.7. The question for statistical inference is whether the observed mean difference of 5.2 points can be attributed to chance as a result of the particular randomization obtained. Cards containing the 20 scores for each group were prepared and submitted to List 12.1, with the results reported in Table 12.3. The two-tailed $P(H_0)$ is .10, but for a one-tailed test $P(H_0)$ is .05, leading to a rejection.

Note that the perfect correlation between mean differences and t values

Table 12.3 Exact Randomization Distribution of *t* for an Experiment

```
EXACT RANDOMIZATION T TEST,  N =  40

INPUT SCORES, GROUP  1
     37    18    46    18    53    26    28    33    30    20
     35    25    18    14    40    19    37    21    21    19
INPUT SCORES, GROUP  2
     36    16    19    25    34    30    29     5    29    18
     30    15    19    12    16    30    19    35    21    16

T TEST FOR OBTAINED SCORES =   1.70

M1 = 27.9    M2 = 22.7    MD =  5.2    S.E. MD =  3.06

ORDERED OUTCOMES FOR RANDOM ARRANGEMENTS FOLLOW
```

SERIAL	MEAN DIF.	T VALUE	SERIAL	MEAN DIF.	T VALUE
1	8.8	3.11	38	2.7	.86
2	7.8	2.68	39	2.7	.86
3	7.5	2.56	40	2.7	.86
4	7.4	2.52	41	2.5	.80
5	6.4	2.14	42	2.5	.80
6	5.5	1.81	43	2.4	.76
7	5.4	1.77	44	2.3	.73
8	5.3	1.74	45	2.3	.73
9	5.3	1.74	46	2.3	.73
10	4.9	1.60	47	2.2	.70
11	4.8	1.56	48	2.2	.70
12	4.6	1.49	49	2.0	.63
13	4.6	1.49	50	1.9	.60
14	4.5	1.46	51	1.9	.60
15	4.4	1.42	52	1.9	.60
16	4.4	1.42	53	1.8	.57
17	4.4	1.42	54	1.8	.57
18	4.4	1.42	55	1.8	.57
19	4.3	1.40	56	1.8	.57
20	4.3	1.40	57	1.8	.57
21	4.3	1.40	58	1.8	.57
22	4.3	1.40	59	1.7	.54
23	4.0	1.29	60	1.6	.51
24	4.0	1.29	61	1.5	.47
25	3.9	1.26	62	1.4	.44
26	3.6	1.16	63	1.4	.44
27	3.5	1.12	64	1.3	.41
28	3.5	1.12	65	1.3	.41
29	3.4	1.09	66	1.2	.38
30	3.3	1.06	67	1.2	.38
31	3.1	.99	68	1.2	.38
32	3.1	.99	69	1.1	.35
33	3.1	.99	70	1.1	.35
34	2.9	.92	71	1.1	.35
35	2.9	.92	72	1.1	.35
36	2.8	.89	73	1.1	.35
37	2.8	.89	74	1.1	.35

Table 12.3 (*continued*)

SERIAL	MEAN DIF.	T VALUE	SERIAL	MEAN DIF.	T VALUE
75	1.0	.32	126	- .8	- .25
76	1.0	.32	127	- .8	- .25
77	1.0	.32	128	- .8	- .25
78	.9	.28	129	- .8	- .25
79	.9	.28	130	- .8	- .25
80	.9	.28	131	- .9	- .28
81	.9	.28	132	-1.0	- .32
82	.8	.25	133	-1.0	- .32
83	.7	.22	134	-1.1	- .35
84	.7	.22	135	-1.1	- .35
85	.7	.22	136	-1.3	- .41
86	.7	.22	137	-1.5	- .47
87	.6	.19	138	-1.5	- .47
88	.6	.19	139	-1.6	- .51
89	.5	.16	140	-1.6	- .51
90	.5	.16	141	-1.7	- .54
91	.5	.16	142	-1.8	- .57
92	.4	.13	143	-1.8	- .57
93	.4	.13	144	-1.8	- .57
94	.4	.13	145	-1.8	- .57
95	.3	.10	146	-1.9	- .60
96	.2	.06	147	-1.9	- .60
97	.1	.03	148	-1.9	- .60
98	.1	.03	149	-2.0	- .63
99	.1	.03	150	-2.0	- .63
100	.1	.03	151	-2.0	- .63
101	.0	.00	152	-2.1	- .67
102	- .1	- .03	153	-2.2	- .70
103	- .2	- .06	154	-2.3	- .73
104	- .2	- .06	155	-2.3	- .73
105	- .3	- .10	156	-2.3	- .73
106	- .3	- .10	157	-2.3	- .73
107	- .3	- .10	158	-2.4	- .76
108	- .3	- .10	159	-2.4	- .76
109	- .3	- .10	160	-2.5	- .80
110	- .3	- .10	161	-2.6	- .83
111	- .3	- .10	162	-2.7	- .86
112	- .4	- .13	163	-2.7	- .86
113	- .4	- .13	164	-2.7	- .86
114	- .5	- .16	165	-2.7	- .86
115	- .5	- .16	166	-2.7	- .86
116	- .5	- .16	167	-2.8	- .89
117	- .5	- .16	168	-3.0	- .96
118	- .5	- .16	169	-3.1	- .99
119	- .6	- .19	170	-3.1	- .99
120	- .6	- .19	171	-3.1	- .99
121	- .6	- .19	172	-3.2	-1.02
122	- .6	- .19	173	-3.3	-1.06
123	- .7	- .22	174	-3.6	-1.16
124	- .7	- .22	175	-3.6	-1.16
125	- .7	- .22	176	-3.7	-1.19

Table 12.3 (*continued*)

SERIAL	MEAN DIF.	T VALUE
177	-3.8	-1.22
178	-3.9	-1.26
179	-3.9	-1.26
180	-3.9	-1.26
181	-3.9	-1.26
182	-4.1	-1.32
183	-4.1	-1.32
184	-4.2	-1.36
185	-4.4	-1.42
186	-4.5	-1.46
187	-4.6	-1.49
188	-4.6	-1.49
189	-4.6	-1.49
190	-4.6	-1.49
191	-5.2	-1.70
192	-5.3	-1.74
193	-5.4	-1.77
194	-5.6	-1.84
195	-5.6	-1.84
196	-5.7	-1.88
197	-7.0	-2.36
198	-7.1	-2.40
199	-7.4	-2.52
200	-8.1	-2.81

suggests that the latter were not really needed to support a statistical inference.

12.2 Summary

Randomization is the most important condition required for statistical inference. When subjects have been randomly assigned to treatments, the distribution of possible outcomes of randomization can be used as a model for statistical inference. The proportion of outcomes as extreme as and more extreme than the actual outcome can be taken as the probability of the actual outcome on the null hypothesis that the particular randomization achieved entirely accounts for the result.

If N is fairly large, there will be too many possible outcomes of randomization to warrant computing all of them. A random sample of the possible outcomes is sufficient to estimate the distribution.

FORTRAN LIST 12.1 takes the obtained scores for two treatment groups and computes and orders a random sample of 200 of the possible randomization outcomes for the experiment. It also returns the t value for the observed outcome, to be referred to the distribution table generated.

This exact randomization test procedure can be generalized to almost any experimental design, thus providing an "all inclusive" nonparametric statistical inference procedure. It is seldom used in practice because of the

SUMMARY

laborious nature of the computing involved. It is useful to gain familiarity with this approach because of the basic statistical principles involved.

EXERCISES AND QUESTIONS

1. Invent a set of data as the observed outcomes of a hypothetical experiment of modest proportions, punch cards for the two groups, submit the cards to List 12.1, and write an interpretation of the results.

```
C        LIST 12.1    EXACT RANDOMIZATION TEST
C
C
C        THIS PROGRAM COMPUTES A REPLACEMENT RANDOM SAMPLE OF 200 POINTS FROM
C     THE POSSIBLE T TEST OUTCOMES OF ASSIGNING N SCORES TO TWO GROUPS IN ALL
C     COMBINATIONS OF N THINGS ASSIGNED N/2 AT A TIME.  N (EVEN) MAY NOT BE
C     MORE THAN 200, AND IS GIVEN BY THE FIRST INPUT CARD (COLS 1-3).  THE NEXT
C     INPUT CARD(S) CONTAIN(S) N/2 SCORES FOR GROUP ONE.  FINAL INPUT CARD(S)
C     CONTAIN(S) N/2 SCORES FOR GROUP TWO.
C
C        THE OUTPUT REPORTS THE ACTUAL T TEST FOR THE OBSERVED MEAN DIFFERENCE,
C     AND THEN GIVES A TABLE OF 200 RANDOMIZATION OUTCOMES FOR THE DATA, ORDERED
C     FROM LEAST T TO LARGEST T.
C        FUNCTION RANDOM(K) IS REQUIRED.
C
         DIMENSION   X(200), NV(200), XS(200), TR(200), XD(200)
C
         READ(5,2)   N
    2    FORMAT (I3)
         WRITE(6,3)  N
         WRITE(7,3)  N
    3    FORMAT (33H1EXACT RANDOMIZATION T TEST, N = I4)
C
         NG = N / 2
         EN = N
         ENG = NG
         NS = 200
         Y = RANDOM(1)
         L = 0
C
         READ (5,4)  (X(J),    J = 1, N)
    4    FORMAT (16F4.0)
C     BOTH FIRST AND SECOND GROUP SCORES ARE READ BY THIS ONE IMPLICIT READ LOOP.
C
         DO 37   I = 1, 2
         JF = (I - 1) * NG + 1
         JL = JF + NG - 1
         WRITE(6,35)    I
         WRITE(7,35)    I
   35    FORMAT (21H0INPUT SCORES, GROUP I3)
         WRITE(6,36)    (X(J),  J = JF, JL)
   37    WRITE(7,36)    (X(J),  J = JF, JL)
   36    FORMAT (20F6.1)
C
         DO 10   J = 1, N
   10    XS(J) = X(J)
         GO TO 24
```

```
      C
   11 WRITE(6,5)    T
      WRITE(7,5)    T
    5 FORMAT (30HOT TEST FOR OBTAINED SCORES = F7.3)
      WRITE(6,6)    XM1, XM2, XMD, SED
      WRITE(7,6)    XM1, XM2, XMD, SED
    6 FORMAT(6HOM1 = F8.3,7H  M2 = F8.3,7H  MD = F8.3,13H   S.E. MD = F8
     C.3)
      WRITE(6,7)
      WRITE(7,7)
    7 FORMAT(48HOORDERED OUTCOMES FOR RANDOM ARRANGEMENTS FOLLOW)
      GO TO 14
      C
   12 TR(L) = T
      XD(L) = XMD
      IF (NS - L)   13, 13, 14
      C
   14 L = L + 1
      DO 15  J = 1, N
   15 NV(J) = 0
      J = 1
   16 K = RANDOM(0) * EN + 1.0
      IF (N - K)    16, 30, 30
   30 IF (NV(K))    17, 17, 16
   17 NV(K) = K
      XS(J) = X(K)
      J = J + 1
      IF (NG - J)   19, 16, 16
   19 I = 0
      DO 22  K = J, N
   20 I = I + 1
   21 IF (NV(I))    22, 22, 20
   22 XS(K) = X(I)
   C  A RANDOM ASSIGNMENT OF SCORES TO GROUPS IS COMPLETED.
   C
   C
   24 SUM1 = 0.0
      SUM2 = 0.0
      SSQ1 = 0.0
      SSQ2 = 0.0
      DO 25  J = 1, NG
      SUM1 = SUM1 + XS(J)
   25 SSQ1 = SSQ1 + XS(J) * XS(J)
      XM1 = SUM1 / ENG
      SSQ1 = SSQ1 - SUM1 * SUM1 / ENG
      K = NG + 1
```

List 12.1 (*continued*)

SUMMARY

```
        DO 26  J = K, N
        SUM2 = SUM2 + XS(J)
 26     SSQ2 = SSQ2 + XS(J) * XS(J)
        XM2 = SUM2 / ENG
        SSQ2 = SSQ2 - SUM2 * SUM2 / ENG
        XMD = XM1 - XM2
        SED = SQRT (((SSQ1 + SSQ2)/(EN - 2.0)) * (2.0 / ENG))
        T = XMD / SED
C
        IF (L)   11, 11, 12
C
C
 13     L = 0
        DO 31  J = 2, NS
        IF (TR(J) - TR(J-1))   31, 31, 32
 32     TEMP = TR(J-1)
        TEMP2 = XD(J-1)
        TR(J-1) = TR(J)
        XD(J-1) = XD(J)
        TR(J) = TEMP
        XD(J) = TEMP2
        L = 1
 31     CONTINUE
        IF (L)   33, 33, 13
 33     WRITE(6,8)
        WRITE(7,8)
 8      FORMAT (27HOSERIAL    MEAN DIF.   T VALUE)
C  T VALUES ARE NOW RANKED FROM LEAST TO LARGEST.
C
        DO 34  J = 1, NS
        WRITE(6,9)    J,  XD(J),  TR(J)
 34     WRITE(7,9)    J,  XD(J),  TR(J)
 9      FORMAT (3X,I4,4X,F7.3,3X,F6.3)
C
C  THIS IS A GOOD OPPORTUNITY TO TEST YOUR ABILITY TO FOLLOW CONDITIONAL
C  TRANSFERS (IF STATEMENTS).  PRETEND YOU ARE THE COMPUTER, AND SEE IF YOU CAN
C  RUN THE PROGRAM THROUGH TO ITS LAST OPERATION, WHICH IS    END
C
        CALL EXIT
        END
```

List 12.1 (*continued*)

CHAPTER THIRTEEN

Statistical Prediction

13.1 The prediction problem

The general concern of this chapter is prediction. It is the ability of scientists to anticipate events that gives them and others confidence in their developed general principles upon which their predictions are based. Statistical prediction will perhaps seem a little different to the student who thinks that predicting is saying what will happen "right on the nose." That is, who thinks that to predict is to be exactly correct. The statistician has a more liberal attitude. He tolerates error in his predictions, but he is also familiar with the nature of that error and makes statements about the expected error in his predictions.

Let us begin with a simple hypothetical example. A psychologist has previously observed the time it takes for white rats from strain A to run through a certain maze. All of the rats do not take exactly the same time. The distribution of observed times is recorded in Table 13.1. Once a sample

Table 13.1 Distribution of Running Times for Strain A Rats

Time in Seconds (T)	Number of Rats (f)	($T \times f$)	($T \times T \times f$)
10	1	10	100
9	6	54	486
8	16	128	1024
7	29	203	1421
6	35	210	1260
5	26	130	650
4	14	56	224
3	7	21	63
2	2	4	8
	$N = 136$	$\sum T = 816$	$\sum T^2 = 5236$

$$M_T = \frac{816}{136} = 6.00$$

$$E = \frac{5236}{136} - 36 = 38.5 - 36 = 2.5$$

$\sqrt{E} = 1.58,$ the standard error of prediction

THE PREDICTION PROBLEM

of A-strain rats has been observed in this fashion the psychologist may want to predict the time needed to run the maze for a rat from strain A who had not previously been observed. Let T_i represent the observed time for rat i to run the maze. G is the time guessed and the guesses will be the same for all rats from strain A, since, for now at least, we have no basis for distinguishing among the members of strain A in order to make a different prediction for each rat. The amount of error involved in making the time estimates will depend upon the value chosen for G. If we choose to define error as the average square difference between guesses and the time finally observed then a familiar equation comes into the picture.

$$E = \frac{\sum (T_i - G)^2}{N} \tag{13.1.1}$$

That is, if we define error as E in the above equation and if we want to find a value for our guess G so as to make our error for N predictions as small as possible, then we have the same equation and the same problem as we had in Part 3.2 of Chapter 3. There we found that if G is chosen to be the mean for the T_i observations, then error E will be as small as possible.

If G is used as the predicted time for all rats of strain A, then almost all of the predictions will be "wrong" in the strict sense, but using G is still the best strategy in the long run because it will produce the smallest average squared error. The amount of error in each prediction will vary. Sometimes it will be large. Usually it will be small. We now need a measure of this variation in error. Formula 13.1.1 was the way we defined error and so that can be our measure of error variance. If the observations are normally distributed about the predicted value G, then the square root of this error variance can be used just as a standard deviation is used with the normal curve table to tell how the predictions will compare with the subsequent observations. This square root of the error variance we can call the standard error of prediction. In Table 13.1 this standard error is 1.58. Therefore it is possible to say that about 68 percent of the observed times will fall within about 3 seconds of the predicted time (plus and minus 1 standard deviation). Is this true for the data in Table 13.1?

Since the value for G is based upon observing T_i for a sample of rats from strain A, G will be an appropriate predicted time only for rats from this strain. If it is of interest to predict for members of other populations, then new samples must be drawn for those other populations in order to obtain estimates of their parameters. Table 13.2 summarizes observations for three strains. Notice that the three sample means are different. It now appears justifiable not to use strain A as a best guess for strains B and C.

Are you now prepared to believe that strain A rats are not a random sample from the same populations represented by samples B and C? How

Table 13.2 Means and Standard Errors for Strains *A*, *B*, *C*, and Total

	Mean Time in Seconds	Standard Error
Strain *A*	6.0	1.58
Strain *B*	8.0	1.32
Strain *C*	10.0	1.41
Combined	8.0	2.15

would a statistician ask that question? In terms of useful differences in making predictions, further study of Table 13.2 should convince you that predictions are better (that is, average errors are smaller) when the strain of the rat is taken into account in making predictions about times needed to run mazes.

In Table 13.2 the mean and standard error for all three strains combined are also presented. Notice that the standard error of prediction here is greater than it is for each of the separate strains. If we had started by predicting for rats regardless of strain we would have discovered that our predictions would be improved by taking strain into account. One of the favorite games of behavioral scientists is seeking ways to reduce this error variance by introducing more information into the predictive system. For example, we could take sex into account here and make different predictions for male and female rats. Table 13.3 summarizes these differences for strain *A*. The standard error of prediction has been reduced by taking sex into account in making these predictions.

In addition to the more genetically determined differences such as strain and sex, prediction can often be improved by using environmental information. In the case of the rats, for example, a knowledge of whether the rat had had any previous experience in maze running may be useful in reducing

Table 13.3 Sex Difference in Strain *A*

Time in Seconds	Number of Males	Number of Females	Combined Frequency
10		1	1
9		6	6
8	2	14	16
7	10	19	29
6	15	20	35
5	19	7	26
4	13	1	14
3	7		7
2	2		2
	$N_m = 68$	$N_f = 68$	$N = 136$
Means (nearest hundredth)	5.12	6.88	6.00
Standard Error	1.38	1.25	1.58

THE PREDICTION PROBLEM

prediction error. Table 13.4 shows the difference between experienced and inexperienced males. Once again we have reduced the error or "unexplained variance," but this time only slightly.

Because there are so many possible sources of differences for any criterion, scientists have begun to use computer programs to help in organizing these differences for the purpose of making predictions. Computer programs have

Table 13.4 Effect of Training on Strain *A* Males

Time in Seconds	Number of Experienced	Number of Inexperienced	Total Number of Males
9			
8		2	2
7	4	6	10
6	5	10	15
5	11	8	19
4	8	5	13
3	4	3	7
2	2		2
N	34	34	68
Means	4.74	5.50	5.12
Standard Error	1.32	1.33	1.38

been developed that produce prediction tree structures. The program will eliminate those branches which are inappropriate because the differences they represent are not sufficiently large to be either statistically significant or of practical predictive importance. Figure 13.1 illustrates the tree produced by such a program for the rat data. The branching is in order of the predictive importance of the various characteristics.

The prediction tree approach can also be illustrated using some of the TALENT data. The "game" might be to find branches which help to predict reading comprehension ability (variable 12). Figure 13.2 summarizes the results of using all the Appendix B cases (both sexes) with the following predictors.

Variable Number		Variable Range
1	School size	1–4
2	Geographic region	1–9
3	Age	15–20
4	Sex	1,2
6	Weight	1–12
7	Type of college planned	1–9
8	Plan college full time	1–5
20	SES	0,1

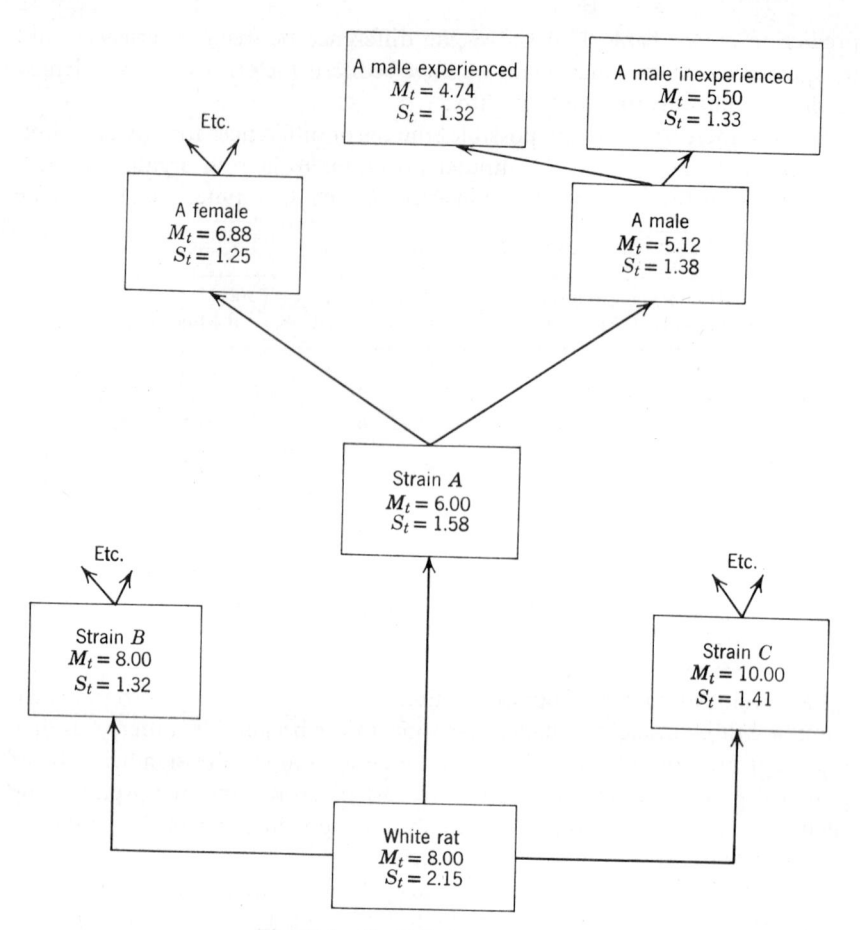

Fig. 13.1 Prediction tree structure.

[Tree diagram showing hierarchical prediction structure:

- A male experienced: $M_t = 4.74$, $S_t = 1.32$
- A male inexperienced: $M_t = 5.50$, $S_t = 1.33$
- A female: $M_t = 6.88$, $S_t = 1.25$
- A male: $M_t = 5.12$, $S_t = 1.38$
- Strain A: $M_t = 6.00$, $S_t = 1.58$
- Strain B: $M_t = 8.00$, $S_t = 1.32$
- Strain C: $M_t = 10.00$, $S_t = 1.41$
- White rat: $M_t = 8.00$, $S_t = 2.15$]

Variable 20 is used here as a dichotomous variable since we told the computer to read only column 63. Thus, SES scores of 100 and above have a "1" and scores below 100 have a score of "0" on variable 20 as used here.

The Figure 13.2 tree tells you that if all you know is that this is a twelfth grader in the U.S.A. in 1960, then your best prediction is a score of 33.7. Your standard error for that prediction is the same as the population estimate of the standard deviation, namely, 9.2. If you also knew that the subject was definitely planning college, or at least almost sure to go (responses 1 and 2 on variable 8), then your prediction would be 38.0 with a standard error of 8.3. If on the question of type of college the subject responded 1, 6, 8, or 9, then your prediction would not be so high (only 32.6); unless the subject was a light weight which would shoot your prediction back up to 38.5.

THE PREDICTION PROBLEM

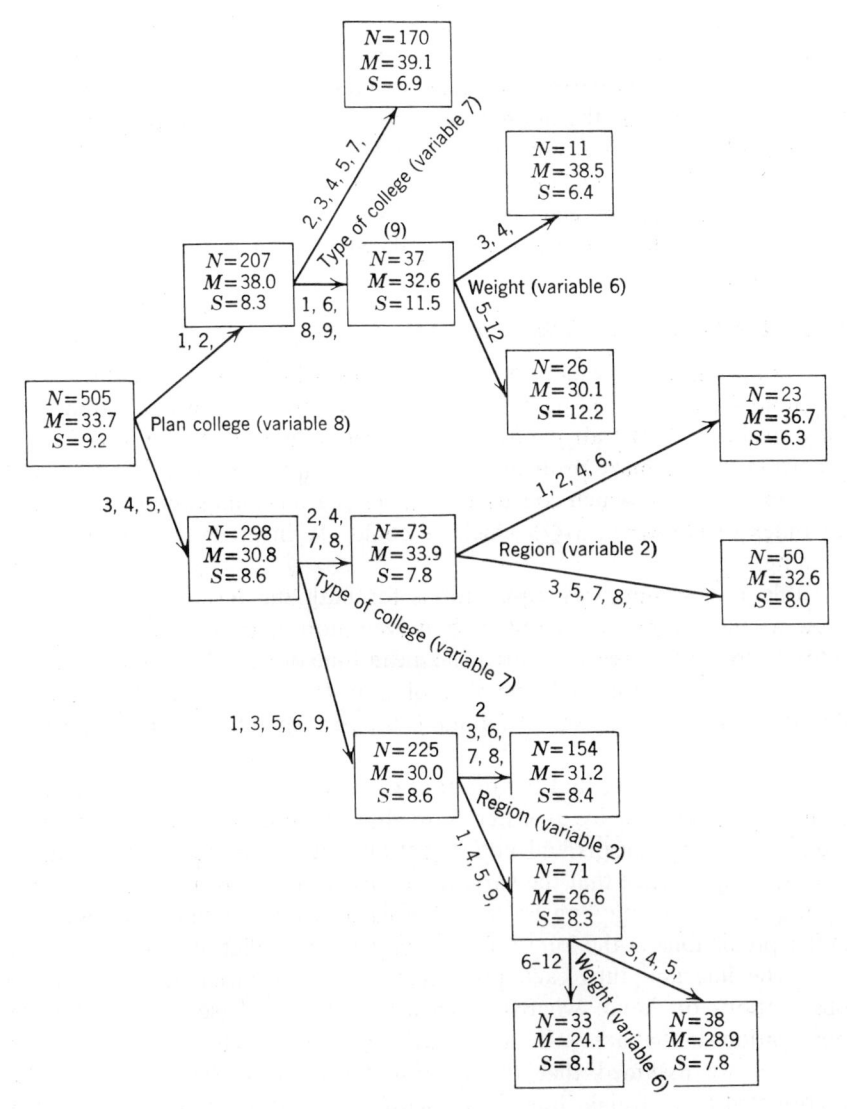

Fig. 13.2 Prediction of reading comprehension.

Your lowest reading comprehension prediction would be for those twelfth graders who were not so sure of college, who responded 1, 3, 5, 6, or 9 to type of college, who were from regions 1, 4, 5, or 9, and who weighed more than 135 pounds. For them you would predict a score of 24.1 with a standard error of 8.1.

Notice that not all of the predictors were found useful (not even sex!). This does not necessarily mean that the unused predictors were unrelated to the criterion. It does mean that other predictors were found more useful at first, and by the time the unused ones may have been useful predictors, the cell sizes (N) became too small to develop a useful split or prediction branch. Even in this illustration some of the N's became a little small to expect confident estimates of cell means. In Section 13.4 you will see a more efficient approach to using variable 20, SES, as a predictor of reading comprehension scores.

13.2 Prediction through regression

In addition to considering characteristics which form categories, the psychologist often uses measurements of traits to assist in prediction. The most useful type of trait is one that is closely related to the criterion to be predicted. In the maze problem above, for example, the psychologist could develop a little test which consisted of actual maze running and then develop an Index of Quickness (IQ) which he could use in predicting future maze running.

Table 13.6 summarizes observations for both the IQ and the criterion. Now we have improved the prediction even more than we did by using the information in the tree structure. The mean time of each IQ group can serve as the prediction for each member of that IQ group, and the standard deviation of each group or IQ column is the corresponding standard error of prediction.

In Table 13.6 notice that the mean of each vertical array (that is, each IQ column) is indicated by an asterisk in that column. Hold a straight edge along these array means and notice that all of them lie nearly in a straight line. We might argue that the variation of array means from the straight line is simply due to sampling error within each array and that we might get better predictions in the long run if we based our predictions on the straight line. The line and thus each prediction would be based upon all of the observations on hand, whereas each array mean is based only upon the observations in that array. In other words, we could develop a new prediction device if we assumed that the relationship between IQ and T can be represented by a straight line. The equation for that line would serve as the basis for predicting T given the value for IQ.

We now need to review a little analytic geometry to find out how straight lines are expressed as equations and how the resulting equation will help in prediction. In Figure 13.3 a straight line is used to represent the relationship between X and Y. If the value of X is 2, move vertically up from 2 on X to the line and then horizontally over from the line to the Y axis. There the value 4 indicates that when $X = 2$ then $Y = 4$. Similarly when $X = 12$

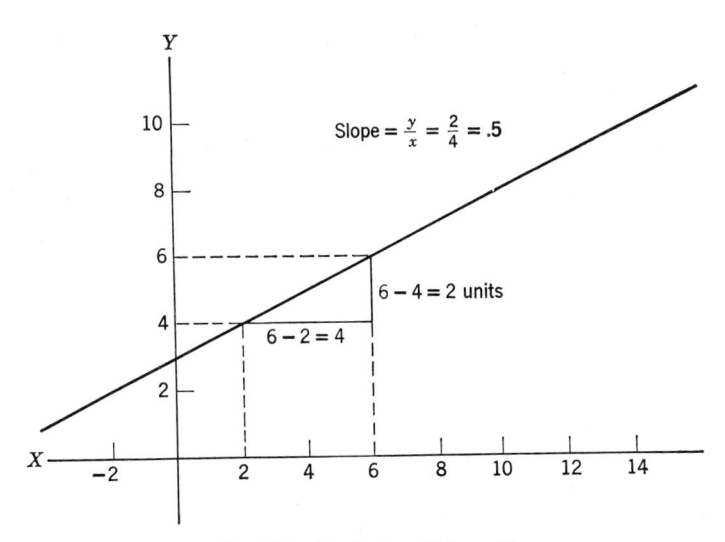

Fig. 13.3 Predicting Y from X.

Table 13.6 Distribution of Times for Different IQ Values (Strain A Rats)

Time in Seconds (T)	\multicolumn{8}{c}{Index for Quickness (IQ)}	
	0	1	2	3	4	5	6	7	Total
10								1	1
9					1	1	1	3*	6
8				1	2	5	5*	3	16
7				4	10	10*	4	1	29
6			1	10*	*13	10	1		35
5		1	4	7	10	4			26
4		3	*4	4	2	1			14
3		3*	2	1	1				7
2		1	1						2
Column Totals		8	12	27	39	31	11	8	136
Mean T		3.50	4.17	5.56	6.0	6.55	7.55	8.5	6.00
Standard Error T		0.86	1.05	1.11	1.17	1.10	0.74	1.41	1.58

Mean $IQ = 4.01$
IQ Standard Deviation $= 1.48$

* Asterisk indicates mean of each vertical array.

then $Y = 9$. In such a simple example the task of determining Y given X can be done graphically. If we knew the equation for this line, we could also go from X to Y algebraically. Geometricians tell us that the general equation for a straight line is $Y = bX + a$, where a is the value of Y at the point where the line intercepts the Y axis, that is, where $X = 0$. The value b is simply the ratio of the height Y to the length X of any right triangle we might choose to draw using the line as the hypotenuse. This b is called the slope of the line. From this we learn that the slope of the line in Figure 13.3 is 0.5.

In such a simple figure it is also possible to determine the value of Y when $X = 0$ by inspection. Since this is seen to be equal to 3 we now have our intercept constant and the equation for the line in Figure 13.3 can now be written: $Y = 0.5X + 3$. If we have a value for X and want to find Y we can do so algebraically by substituting the value for X in this equation. Thus for $X = 6$, $Y = 0.5(6) + 3 = 6$. This is confirmed by inspecting Figure 13.3.

Now that we have a general equation for a straight line we can go back to our rat problem and see if we can figure out an equation for predicting T given G. Just by inspecting Table 13.6 we can see approximately what the line's equation ought to look like by once again lining up the array means. The slope b looks like seven units over and five units up, and 5/7 is about 0.7. When $IQ = 0$, T is approximately 3 (our guess for a). These estimated values for a and b can be substituted into our general equation for a line: $T = (.7)(IQ) + 3$. Thus for an IQ of 6 we would predict a time of about 7 seconds. The algebraic "answer" of 7.2 implies a degree of precision which is unwarranted when using these graphical procedures.

Since this graphical method is very imprecise we need to figure out a procedure for finding a prediction line which gives as little error in prediction as possible. Of course there has to be some error because not all of the observed points lie on a straight line.

In order to simplify the algebra to be presented below without changing the nature of our problem, the observations recorded in Table 13.6 have been converted to deviation scores in Table 13.7. In the case of deviation scores the a term of the straight line equation is eliminated because the line goes through the place where the T axis intercepts the IQ axis. That is, the line goes through the origin in this case. This seems reasonable since if the rat in question had an average IQ we would predict an average time T, so if $x = 0$ then $y = 0$ also. Our problem now is to determine the slope of this line which goes through the origin.

One way to begin is to decide upon a criterion for locating the line. We might, for example, decide that a line is desired which produces the smallest vertical distance between any one point and the line. A problem with this

PREDICTION THROUGH REGRESSION

Table 13.7 Joint Distribution of T and IQ in Deviation Form

Deviation Time	Deviation IQ						
	−3	−2	−1	0	1	2	3
4							1
3				1	1	1	3
2			1	2	5	5	3
1			4	10	10	4	1
0		1	10	13	10	1	
−1	1	4	7	10	4		
−2	3	4	4	2	1		
−3	3	2	1	1			
−4	1	1					

approach is that one point of observation could pull the line quite a distance from most of the other points. A better criterion would be to find a line which would minimize the average prediction error. Error can again be defined best as the square of the difference between our prediction (T') and the observed time (T). Recall that we square errors so that negative errors don't cancel positive ones and to emphasize extreme errors.

Dealing with the algebra in terms of more general notation, using X and Y instead of IQ and T, respectively, we can begin by defining our problem as finding a line which would minimize the mean of squared differences between prediction (Y') and observation (Y):

$$\frac{\sum (Y - Y')^2}{N} \quad \text{or in deviation terms:} \quad \frac{\sum (y - y')^2}{N}$$

In these terms our straight line equation becomes $y' = bx$. Since b times x is equal to y' we can substitute bx for y' in the expression we wish to minimize. Setting this new expression equal to, say E, we have:

$$E = \frac{\sum (y - bx)^2}{N}$$

Our problem now is to find a value for b which would make E, the predictor error, as small as it can get.

To solve this minimization problem a branch of mathematics known as differential calculus is needed. Calculus provides a set of rules for manipulating equations for purposes like solving for minimum values in equations such as ours. We do not expect all students to fully understand the steps involved in the derivation which follows, but we do urge that you see why these steps are being taken, and we also hope that by presenting a few derivations like this you will begin to appreciate that behind all of the statistical procedures in this book there is a sound mathematical discipline upon which these procedures are based. Here goes! We begin with the equation:

$$E = \frac{\sum (y - bx)^2}{N}$$

The "magic" step is to differentiate E with respect to b. The result of this differentiation process is:

$$\frac{dE}{db} = \frac{2 \sum (y - bx)(x)}{N}$$

This derivative of E with respect to b is set equal to 0 in order to obtain a value for b which minimizes E:

$$0 = \frac{2 \sum (y - bx)(x)}{N}$$

This equation can be simplified on the right-hand side by multiplying bx and y by x, just as the parentheses indicate. We can also divide both sides of the equation by the constant 2, and multiply both sides by N, resulting in:

$$0 = \sum (xy - bx^2)$$

The right-hand side of the equation can be broken up into two terms, distributing the summation sign $\sum$ and using the principle that the slope b we are seeking can be brought in front of the summation sign because a constant times the sum of x^2's is the same as summing the constant times each of the x^2's (summation rule III, Part 3.2). Thus we now have:

$$0 = \sum xy - b \sum x^2$$

Adding $b \sum x^2$ to both sides of the equation:

$$b \sum x^2 = \sum xy$$

Dividing both sides by $\sum x^2$, a means of computing b is finally obtained in terms of deviation values for x and y:

$$b = \frac{\sum xy}{\sum x^2} \qquad (13.2.1)$$

Thus the desired prediction equation is

$$y' = bx = \left(\frac{\sum xy}{\sum x^2} \right) \cdot x$$

THE STANDARD ERROR OF ESTIMATE

The data of Table 13.7 yield a value for b of .77, so we now have the slope constant for predicting deviations from the mean of T given deviations from the mean IQ.

Algebraic exercises similar to the above also make it possible to derive equations for determining a and b in raw score form. The derived results can be expressed as follows:

$$b_{yx} = \frac{\sum XY - [(\sum X)(\sum Y)/N]}{\sum X^2 - [(\sum X)^2/N]} \qquad (13.2.2)$$

$$a_{yx} = \frac{\sum Y - b_{yx} \sum X}{N} \qquad (13.2.3)$$

Applying these formulas for b and a to the raw score data of Table 13.6 you obtain $b = .77$ and $a = 2.91$. Thus the prediction equation is: $T' = (.77) \times (IQ) + (2.91)$. For $IQ = 6$, the predicted T' would be 7.64. Comparing this algebraically derived result of 7.64 to the 7 obtained by graphical means we see that both agree as well as we could expect. However, the algebra was necessary because "real life" data are not so neat as the data for this specially contrived hypothetical example. Later in this chapter we shall apply what has been learned here to some actual data.

13.3 The standard error of estimate

Now that we have a method of determining the prediction equation from our data by assuming that the relationship between T and IQ (or X and Y) is linear, we also need a method of describing the error expected when using this prediction system with future strain A rats. Up to this point the predicted value has been the sample mean of a defined population, such as trained, male, strain-A rats, and the prediction error has been described in terms of the variance or standard deviation about that mean. When we used the mean of the IQ arrays as the prediction for each member of the array, the standard deviation of the vertical array served as the standard error of prediction. But what should we use in describing the error variation about the prediction line?

Previously we had decided upon a measure of error which was our original definition of error in determining that the mean was our best guess, if error, as defined, was to be minimized. Doing this for the prediction line case we get a very similar equation for error, using $S_{y \cdot x}$ to denote the standard error of estimate when predicting Y from a knowledge of X.

$$S_{y \cdot x}^2 = \frac{\sum (Y - Y')^2}{N} \qquad (13.3.1)$$

This is not only how we defined error in arriving at the prediction line equation but the equation is similar to (13.1.1) when you remember that Y' is the predicted value as was our guess G. The difference now is that Y' may be different for each prediction, depending upon the value of X.

Equation 13.3.1 is not too practical a computational formula, however, since it requires computing the estimated score Y' for each observation in the sample. Fortunately, once again we can call upon algebra to transform an impractical definition formula into a more convenient computational formula. Beginning with Equation 13.3.1, and substituting $(bX + a)$ for Y', the result is:

$$S_{y \cdot x} = S_y \sqrt{1 - \left(b_{yx} \frac{S_x}{S_y}\right)^2} \qquad (13.3.2)$$

Since we have the slope $b = .77$, $S_T = 1.58$, and $S_{IQ} = 1.48$, we can readily obtain the standard error for predicting T from IQ.

$$S_{T \cdot IQ} = 1.58 \sqrt{1 - \left(0.77 \frac{1.48}{1.58}\right)^2} = 1.09$$

Notice how this standard error of estimate is similar to the values for the array standard deviations reported in Table 13.6. What we have obtained in $S_{T \cdot IQ}$ is a sort of average of array values, assuming that in the population all of the array standard deviations are equal. Because this is such an important assumption in regression theory it has a very important sounding name. It is called the assumption of homoscedasticity.

If we further assume that the distributions of observations within arrays are normally distributed about the regression line, we now can establish an interval about our predicted value within which a certain proportion of the observed values will lie.

Applying these two new assumptions to our rat problem, we can now ask questions such as the following. What proportion of strain A rats with $IQ = 6$ could be expected to have a smaller T than the strain A mean? Figure 13.4 may assist in thinking about this question.

A useful device in two variable (bivariate) problems is to pictorially describe that bivariate distribution as an ellipse, the perimeter of which approximately outlines the observed swarm of points of that distribution. Looking at Figure 13.4 and Tables 13.6 and 13.7 notice how the ellipse does in fact do this. Notice also in Table 13.6 that the large numbers are at the center with frequencies thinning out at the edges. To take this varying frequency or density into account we must visualize the ellipse drawn there as actually representing a three-dimensional figure similar to a watermelon sliced in half the long way with the flat side down. Now if we take a slice or cross section out of this watermelon at $IQ = 6$ we expect, with the bivariate

THE STANDARD ERROR OF ESTIMATE 211

normal assumption, to find the slice looking like the cross section illustrated in Figure 13.4.

The predicted mean of this $IQ = 6$ array is $T' = 7.6$, and the standard error of estimate $S_{T \cdot IQ}$ for any vertical array in this bivariate distribution was found to be 1.09. Using these values, $T = 5$ can be converted to a normal deviate as follows:

$$\frac{T - T'}{S_{T \cdot IQ}} = z = \frac{5.0 - 7.6}{1.09} = 2.47$$

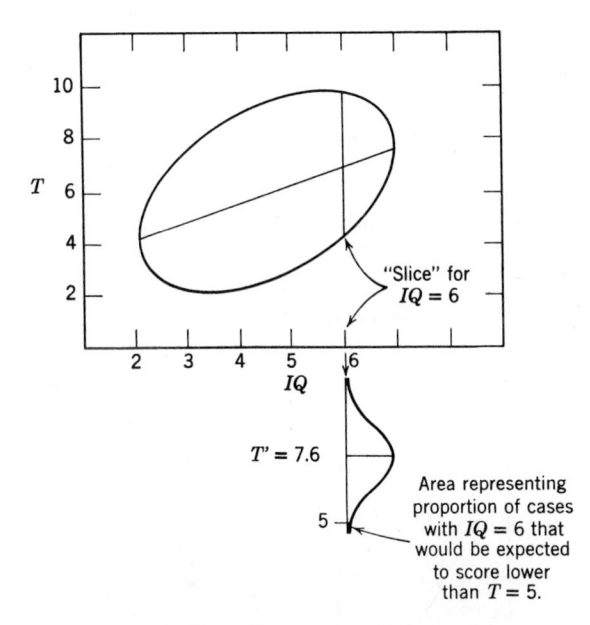

Fig. 13.4 Prediction within a bivariate array.

Entering the normal curve table we find that .001 is the proportion of cases which would be expected to lie further from the array mean of 7.6 than does $T = 5$. Thus should the investigator want to choose rats for his experiment that would tend to take longer than 5 seconds to run the maze, if he used rats whose IQ was 6, about 1 in 1000 would take less than 5 seconds to run the maze.

It is important to note that the regression line that is used to predict Y from X is not the same line that is used to predict X from Y. The reason for this is illustrated in Figure 13.5. There the same ellipse is depicted twice. In the top figure the regression line with slope $b_{yx} = y/x$ is passing through or near the means of the vertical arrays. In the bottom figure the line is passing through or near the means of the horizontal arrays, and the slope of that line is $b_{xy} = x/y$. Of course both lines pass through the point where M_x and M_y intersect.

Notice also that in regression theory no assumptions are made about the nature of the predictor variable. When we are predicting Y from X, X need not be a normal variate, or even ordinal for that matter. The only assumption

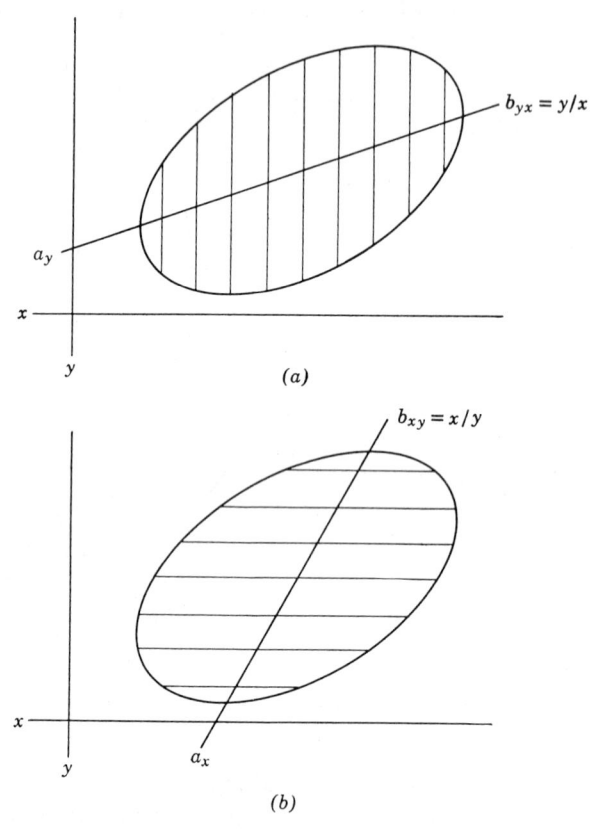

Fig. 13.5 Comparison of b_{yx} and b_{xy}. (*a*) Regression line for predicting y from x, passing through centers of vertical arrays. (*b*) Regression line for predicting x from y, passing through centers of horizontal arrays.

we have made which involves X is that the relationship between X and Y is approximated by a straight line. In fact, as far as the actual prediction problem is concerned, that is also the only assumption involving the criterion Y. It is only when we wish to make statements about the nature of the prediction error that the assumption of homoscedasticity (array standard

SUMMARY

deviations are equal) and the assumption that Y is normally distributed within arrays are necessary.

13.4 Example using Project TALENT data

To further illustrate some of the statistical concepts introduced in this chapter, we can once again turn to the Appendix B data. For example, an investigator may be interested in knowing how well he can predict the reading comprehension ability (Y) of a male student given an index of the socioeconomic environment (X) of his home. The program List 14.2 enables us to do regression analyses quite easily. These results can then be compared to the prediction tree generated for this same criterion in Table 13.6.

The results of running variables 12 and 20 for the males of Appendix B.2 are as follows:

Reading Comprehension (Y)	Socioeconomic Status (X)
$M_y = 33.5$	$M_x = 98.5$
$S^2 = 90.5$	$S^2 = 89.2$
$S = 9.5$	$S = 9.4$

Regression equation to predict Y from X

$$Y_1 = (.371) \cdot X - 2.967$$

That is, the regression slope for predicting reading comprehension from socioeconomic status is 0.371 and the intercept constant is -2.967. If someone had a socioeconomic status score of 100, what would be his predicted reading score? Substituting 100 for X in the equation we get $(37.1 - 2.967)$ or about 34.2. If you use the mean of X as the predictor, the prediction will be the mean of Y.

The standard error of estimate is also printed out by the program. In this case, $S_{Y \cdot X} = 8.8$, which is not much of a reduction of the standard deviation of 9.5. Therefore, the family socioeconomic environment does not appear to be a powerful predictor of reading ability. In the next chapter we shall learn how to interpret some of the other results which are printed out by List 14.2. These other results will help us make inferences about the relationship between X and Y.

13.5 Summary

Regression analysis assumes that the relationship between a predictor trait and a criterion trait can be represented by a straight line. Differential calculus is employed to derive the equation for a line that will lead to a minimization of the error variance

$$\left. \frac{\sum (y - y')^2}{N} \right|_{min}$$

The resulting equation is of the form $y' = bx$ where

$$b = \frac{\sum xy}{\sum x^2}$$

This b is the regression coefficient for the regression of y (the criterion) on x (the predictor).

The standard error of estimate

$$S_{y \cdot x} = S_y \sqrt{1 - \left[b\left(\frac{S_x}{S_y}\right)\right]^2}$$

is the square root of the minimized error variance.

Statistical inference based on the standard error of estimate is possible if we are willing to assume that the distributions of y for all the x arrays are normal and have a common standard deviation. With these assumptions, we can create a confidence interval around y', for example,

$$y' \pm 2S_{y \cdot x}$$

and assert with 95 percent confidence that y exists within these limits.

EXERCISES AND QUESTIONS

1. Using List 14.2 and Appendix B.3 data, compute the regression equation for prediction Reading Comprehension (variable 12) from Socioeconomic Status (variable 20). How does this equation for females compare with the example in Section 13.4 for males?

2. In the prediction tree example of Figure 13.2, sex was not found to be a useful predictor of reading comprehension scores. How would you decide whether or not there was a significant relationship between sex and reading comprehension in the Project TALENT data?

CHAPTER FOURTEEN

Correlation

14.1 A measure of relationship

In addition to an interest in predicting one variable from a knowledge of another, scientists are also interested in talking about the extent to which one variable is related to another. For example, how strong is the relationship between amount of father's education and the student's mathematics achievement scores? How does that relationship compare with that found between the amount of mother's education and the student's mathematics achievement? Here we are not necessarily interested in predicting one variable from the other; rather, the interest is in the magnitude or degree of association between two variables.

An inspection of Figure 14.1 might suggest that the slope b of the regression line might serve as this measure. In part A of Figure 14.1 the narrow ellipse suggests that the observations are fairly close to the regression line, thus the values of Y are closely related to values of X. As the ellipse becomes "fatter" the slope lowers as seen in comparing part A with part B. In part C of Figure 14.1 the lack of any relationship between X and Y suggests that the mean of Y is the best prediction of Y regardless of X, indicating that knowing X tells nothing about what to expect in Y. Thus it appears that the larger the slope b, the greater the relationship.

One problem with using the slope of the regression line as the desired measure of relationship is illustrated in Figure 14.2. There we see two ellipses of the same size and shape yet the slopes are quite different. This is because the X axis is on a different scale in part A than it is in part B. This scale difference is reflected in the two different standard deviations for parts A and B.

Another difficulty with using slope b as the measure of relationship between two variables is that two different values of b are available for the same pair of variables, b_{yx} and b_{xy}. It is highly desirable to have a single measure of the relationship between any two variables given a set of sample observations.

One approach to solving this problem might be to standardize both

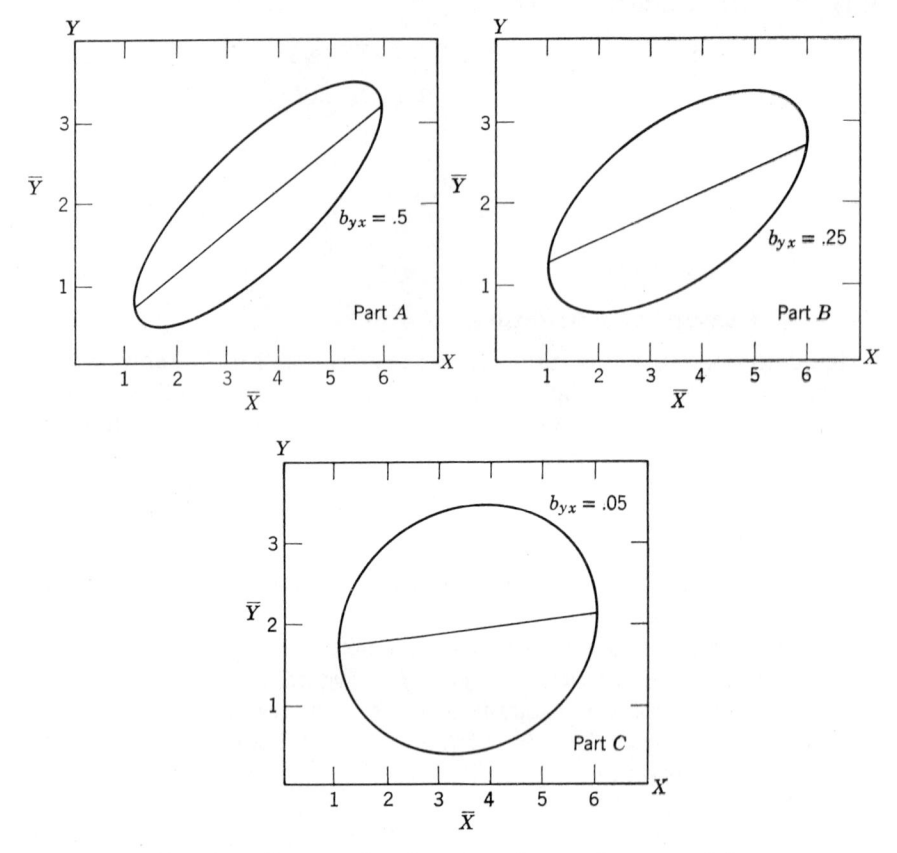

Fig. 14.1 The regression slope depends upon the extent of the relationship for given M_x and S_x.

measures. If they both have a standard deviation of 1, the scale problem of Figure 14.2 is eliminated. Now we need to see what this standardization does for the single measure problem. That is, is the slope the same for the Y to X prediction and for X to Y when the two measures have the same standard deviation? The answer to that question is *yes*. Perhaps Figure 14.3 will help you to see why this is so.

Although the two regression lines in Figure 14.3 appear to have different slopes, this apparent difference is eliminated when you remember that the predictor variable must serve as the base for the right triangle from which the slope is determined. If Figure 14.3, part *A* is redrawn by interchanging axes it will look just like part *B*. Notice that in order to obtain this measure of relationship free from scale factors one additional assumption was

THE CORRELATION COEFFICIENT

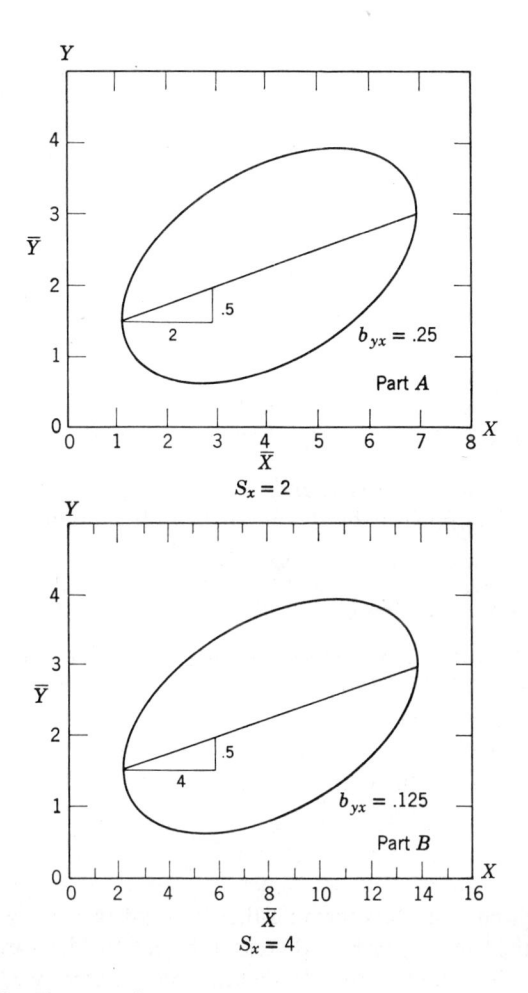

Fig. 14.2 The regression line slope depends upon S_x for a given ellipse.

needed. This was the assumption that the predictor X was also a variable for which standardized scores with a mean of 0.0 and a standard deviation of 1.00 could be appropriately computed. Up to this point no assumptions about the nature of this predictor variable X were necessary.

14.2 The correlation coefficient

Going back to the rat data of Table 13.6, if both T and IQ are converted to z scores, then the slope for these new scores can be computed quite easily by simplifying the original equation for determining the slope (13.2.1). In the

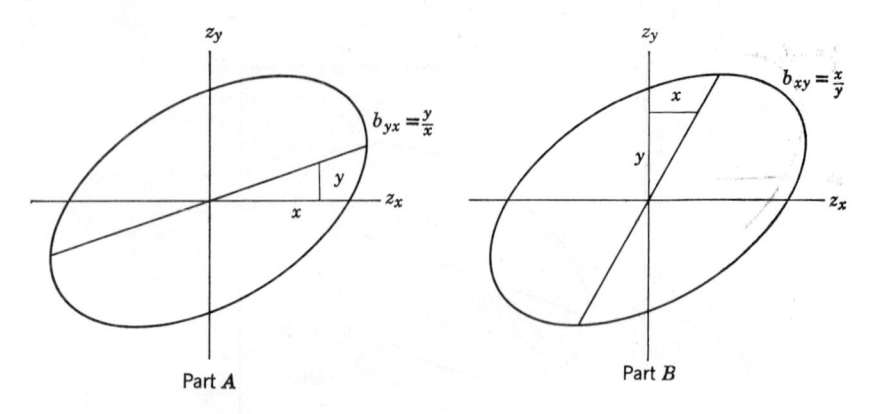

Fig. 14.3 The regression lines in standard score units.

Chapter 13 derivation it was found that $b_{yx} = \sum xy / \sum x^2$. Dividing both numerator and denominator by the number of observations, N, we get:

$$b_{yx} = \frac{\sum xy}{N} \Big/ \frac{\sum x^2}{N}$$

Here the denominator $\sum x^2/N$ is the sample variance, which in this case of standard scores would be 1.00, so the equation can be simplified to become the average cross product of the standard scores of x and y. Let us denote this slope as r when it is computed from standard scores, and call this measure of relationship the correlation coefficient.

$$r = \frac{\sum z_x z_y}{N} = \frac{\sum z_y z_x}{N} \qquad (14.2.1)$$

It can also be seen from this formula that it is not relevant whether we are talking about the slope r_{xy} or r_{yx}, whereas in Equation 13.2.1 there was either $\sum x^2$ or $\sum y^2$ in the denominator, depending upon whether we were considering b_{yx} or b_{xy}, respectively.

Applying Equation 14.2.1 to the data obtained from Table 13.6 converted to standard scores, the resulting $r = .72$. Although Equation 14.1.1 is simple enough, the task of converting each observation pair to standard scores is quite laborious, so once again the need for more efficient computing procedures arises.

It can be shown algebraically that the already accumulated cross products of deviation scores can be converted to the equivalent in standard scores by dividing by the two sample standard deviations. That is:

$$r = \frac{\sum xy}{NS_x S_y} = \frac{\sum z_x z_y}{N} \qquad (14.2.2)$$

THE BIVARIATE NORMAL DISTRIBUTION

Thus, instead of converting each deviation score to a standard score, we can convert the accumulated cross products of deviations from M_x and M_y to the equivalent accumulation in standard scores.

A drawback of Equation 14.2.2 is that it requires computing deviation scores. This can be eliminated by substituting $(X - M_x)$ for x and $(Y - M_y)$ for y in Equation 14.2.2 and further reducing to a useful computing formula in which only raw score accumulations are necessary.

$$r = \frac{\sum z_x z_y}{N} = \frac{\sum xy}{NS_x S_y} = \frac{N \sum XY - \sum X \sum Y}{\sqrt{N \sum X^2 - (\sum X)^2} \sqrt{N \sum Y^2 - (\sum Y)^2}} \quad (14.2.3)$$

In the raw score formula for r, two square roots are indicated. This could be further simplified by obtaining the single square root of the product $[N \sum X^2 - (\sum X)^2][N \sum Y^2 - (\sum Y)^2]$. However, the advantages of Equation 14.2.3 can be seen when you observe that the sample standard deviations can be obtained from $\frac{1}{N}\sqrt{N \sum X^2 - (\sum X)^2}$. Therefore, by obtaining the accumulations N, $\sum XY$, $\sum X^2$, $\sum Y^2$, $\sum X$, and $\sum Y$ we can easily obtain the means for X and Y, their standard deviations, and our new measure of relationship, the correlation coefficient r.

14.3 The bivariate normal distribution

The nature of the relationship between two variables can be approached from either the point of view of regression theory or correlation theory. In this text we begin with regression theory in Chapter 13 because it seems simpler. Certainly fewer assumptions are involved. But the student of statistics should also be aware of the nature and existence of correlation theory, since it plays such a major role as the student goes on to advanced work in statistics.

If we are willing to assume that both X and Y are random normal variates, their bivariate distribution can be described by the equation:

$$g(X, Y) = \frac{N}{2\pi\sigma_x\sigma_y\sqrt{1-\rho^2}} e\left[\frac{-1}{2(1-\rho^2)}\left(\frac{x^2}{\sigma_x^2} + \frac{y^2}{\sigma_y^2} - \frac{2\rho xy}{\sigma_x \sigma_y}\right)\right] \quad (14.3.1)$$

This equation should be compared to Equation 7.1.3 for the normal curve. From the normal curve equation the height of the ordinate, $g(X)$, at a given point X_i can be obtained if the two parameters, μ_x and σ_x are known for variable X. Now in Equation 14.3.1 there are two random variables, X and Y, and with the additional parameters μ_y, σ_y, ρ_{xy}, it is possible to determine the density in the bivariate distribution for any point defined by the two coordinates, X and Y. In this equation, ρ (rho) is the population correlation coefficient for which r is a sample estimate. Using this equation, for example,

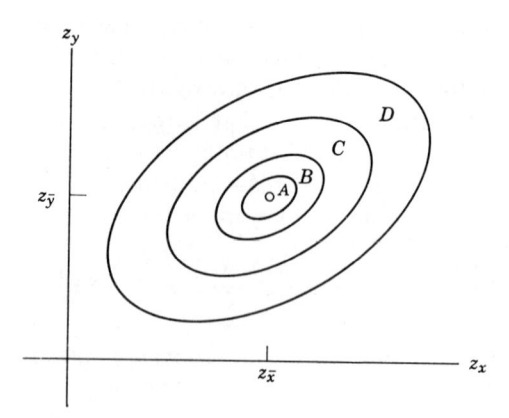

Fig. 14.4 Centours.

we can define ellipses which "trace" points of equal density. Figure 14.4 illustrates this for several different ellipses. Ellipse A is the locus of points of very high density, whereas ellipse D would represent very low density. Low density means that there are relatively few cases with those particular score combinations.

These ellipses have also been called *centours* (*cen*tile con*tours*), because they can be talked about in terms of the proportion of cases which lie further from the center of the ellipse than do points which make up the ellipse. The center of this ellipse is called the centroid, and is the point where $\bar{X}$ and $\bar{Y}$ intersect. In Figure 14.4, ellipse D might represent the centour beyond which only 10 percent of the (X, Y) points lie, and A the centour beyond which 90 percent of the observed points lie. Equation 14.3.1 is used to define such ellipses of equal density.

Another application of Equation 14.3.1 is in the problem of classifying

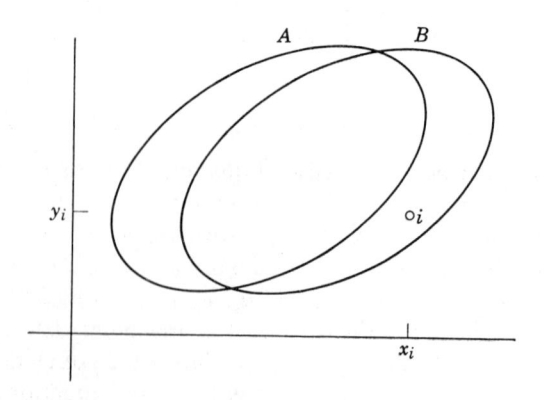

Fig. 14.5 Predicting group membership.

individuals, given their test scores on variables X and Y. In Figure 14.5 the bivariate distributions for two populations, A and B (for example, males and females), are shown, indicating that the two populations tend to occupy different regions of this space defined by axes X and Y. Substituting the parameters for population A into Equation 14.3.2, the density of A's at point (X_i, Y_i) can be found. Similarly, the density of B's at that same point can also be computed. If d_A and d_B are these densities, then the proportion of A's at that point is $d_A/(d_A + d_B)$. It should be clear that

$$\frac{d_A}{d_A + d_B} + \frac{d_B}{d_A + d_B} = 1.00$$

The proportion of A's at (X_i, Y_i) is more properly considered as the probability that individual i (whose population membership is unknown) belongs to

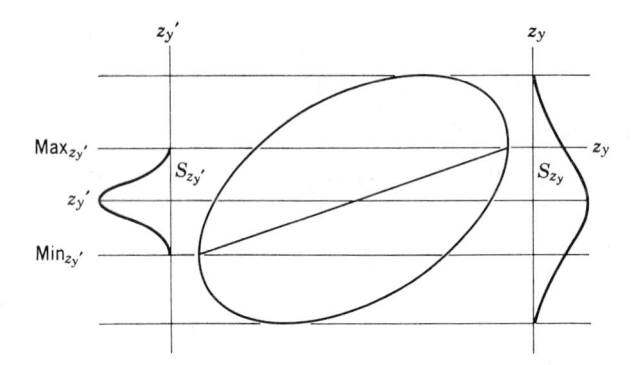

Fig. 14.6 Comparison of observed and predicted distributions z_y.

the A population. Students interested in pursuing this classification problem further might wish to begin with Cooley and Lohnes (1962), Chapter 7.

14.4 Other interpretations of the correlation coefficient

The correlation coefficient r is one of those concepts which continues to grow as we have experience with it. There are many ways of viewing r, in addition to its being the slope of the regression line for two standardized variables. We have also seen r as an estimator of a necessary parameter, ρ, of the bivariate normal Equation 14.3.1. In this section we will look at r from other vantage points.

In Figure 14.6 the standardized bivariate distribution between X and Y is outlined by use of one of the outer ellipses. Also there are distributions for both the observed z_y and the predicted z_y'. Notice that as the slope r changes, the range from maximum z_y' to minimum z_y' also changes, as does $S_{z_y'}$, the standard deviation of the predicted scores. As r approaches zero, so does

S_{z_y}'. Only when $r = 1.00$ is the variation in estimated scores as great as the variation in observed scores. Another way of viewing r, then, is in terms of the extent to which the relationship between X and Y allows us to account for the variation in Y by the predicted value Y'. It turns out that the square of r is the proportion of the variance in Y which can be predicted from the variation in X. Algebraically this is seen as follows, starting with the variance of predicted scores as:

$$S^2_{z_y'} = \frac{\sum (z_y')^2}{N} \quad \text{where} \quad z_y' = rz_x$$

Substituting the latter for z_y' we obtain

$$S_{z_y'} = \frac{\sum (rz_x)^2}{N}$$

and the constant r^2 can be removed from the summation (summation rule III, Part 3.2) giving

$$S^2_{z_y'} = \frac{r^2 \sum z_x^2}{N}$$

Noting that

$$\frac{\sum z_x^2}{N} = 1.00$$

we have finally the variance of the predicted scores as the square of the slope of the prediction line:

$$S^2_{z_y'} = r^2 \qquad (14.4.1)$$

Since the variance of z_y is 1.00, it is therefore possible to speak of r^2 as the proportion of the observed variance in z_y which is predictable from z_x.

14.5 Tests of inference involving *r*

Now that we have seen how to compute a correlation coefficient and have considered some of its properties, it seems reasonable to compute some for real live data. At the end of this chapter is a computer program (List 14.2) for doing just that. Using this program with the Project TALENT data provided in Appendix B.2 we can compute, for example, the correlation between mechanical reasoning ability and interest in the physical sciences for boys. Since the data there are measures on a random sample of grade 12 males in high school in the spring of 1960, the computed correlation of .35 can be considered an estimate of rho for that population.

Now the sample size here is 234 males. The correlation obtained on this sample is certainly just an estimate when you figure that there were about 1,000,000 boys in the grade 12 population that year. The exact *r* we compute

TESTS OF INFERENCE INVOLVING r

from a sample of this size depends upon which 234 pairs of scores are sampled from this very large population. By just happening to select a particular sample it is possible to find a sample r to be large even when the population ρ is equal to zero.

Now that we have had considerable experience with Monte Carlo techniques we can design a study to estimate the sampling distribution of correlation coefficients when $\rho = 0$. Comparing our observed r to the distribution of r's computed from 200 samples of 234 subjects each, we can decide whether our r of .35 is unusually large or not. FORTRAN LIST 14.1 provides such a program. Using $N = 234$ and $RF = .00000023$ in the parameter card the distribution of the resulting 200 r's listed in Table 14.1 was obtained. There we see that all of the computed r's based on 200 random samples from a population where $\rho = 0$ fell between the intervals $-.16$ and $+.18$. Since our observed correlation was .35 we can be rather sure this was not simply the result of chance associations.

Table 14.1 Sampling Distribution of r for $N = 234$ (for $\rho = 0.00$)

Class	Lower Limit	Observed f	Cumulative f
1	$-.18$	0	0
2	$-.16$	2	2
3	$-.14$	3	5
4	$-.12$	3	8
5	$-.10$	11	19
6	$-.08$	14	33
7	$-.06$	25	58
8	$-.04$	20	78
9	$-.02$	20	98
10	$+.00$	31	129
11	.02	25	154
12	.04	20	174
13	.06	10	184
14	.08	3	187
15	.10	7	194
16	.12	2	196
17	.14	2	198
18	.16	0	198
19	.18	2	200
20	.19	0	200

Using smaller samples a quite different result is obtained. Using only 10 subjects correlations can get as large as or larger than .55 about 5 percent of the time when $\rho = 0.0$.

Now to eliminate the necessity of having to compute a Monte Carlo distribution of r every time we are interested in asking whether the population value ρ is probably different from zero, the distribution of r can be converted to the distribution of t, for which tables are readily available. The formula is

$$t = \sqrt{\frac{r^2(N-2)}{1-r^2}}$$

This t-test assumes that the population distribution is bivariate normal and that $\rho = 0.0$. List 14.1 is a program to help convince you that this procedure does in fact produce a distribution which follows the theoretical t distribution, with $N - 2$ degrees of freedom.

Table 14.2 Empirical t_8 for H_0: $\rho = 0$ and Goodness of Fit to Theoretical t_8

Class	Lower Limit	Upper Limit	Observed f	Expected f	$\dfrac{(O-E)^2}{E}$
1	.000	.140	21	20	.050
2	.141	.261	20	20	.000
3	.262	.398	27	20	2.450
4	.399	.545	17	20	.450
5	.546	.705	23	20	.450
6	.706	.888	21	20	.050
7	.889	1.107	19	20	.050
8	1.108	1.396	18	20	.200
9	1.397	1.859	19	20	.050
10	1.860	9.000	15	20	1.250

$\chi_9^2 = 5.00$, $.10 < P < .90$

In Table 14.2 are presented the results of conducting 200 such t-tests based upon samples of 10 score pairs each from independent (uncorrelated) random normal deviate distributions. The upper and lower limits for the distribution program were chosen so that the expected frequency would be 20, assuming that

$$t = \sqrt{\frac{r^2(N-2)}{1-r^2}}$$

in fact follows the t distribution. As can be seen from the resulting chi-square, there is no reason to believe that it does not do so.

14.6 The correlation matrix

One of the most frequent ways in which a set of variables is described is in terms of all possible correlations between pairs of variables within the set. When coefficients are arranged in a two-dimensional array in which corresponding rows and columns designate certain variables, this is known as a correlation matrix. In Table 14.3 we show the correlation matrix for six of the Table B.1 variables for TALENT males ($N = 234$).

Table 14.3 Correlation Matrix (Based on 234 Project TALENT Grade 12 Males)

		1	2	3	4	5	6
	Means	2.90	156.50	82.78	13.57	6.88	21.30
	Standard Deviation	1.56	35.16	12.55	3.58	3.01	9.19
1	SIB 301	1.00	−.49	−.34	−.29	−.23	−.47
2	R-190	−.49	1.00	.63	.60	.05	.57
3	R-230	−.34	.63	1.00	.34	.14	.33
4	R-270	−.29	.60	.34	1.00	.00	.35
5	R-601	−.23	.05	.14	.00	1.00	.21
6	P-701	−.47	.57	.33	.35	.21	1.00

Notice that among the fifteen different correlations presented there, some are significant in the negative direction, some in the positive direction, and some are not significantly different from zero. In fact, the observed correlation between (1) mechanical reasoning ability and (2) sociability is 0.00. Can you explain why the correlations with college plans (SIB 301) are all negative? Refer back to Table B.1 for the coding scheme for help on the answer.

The correlation matrix is a good thing to introduce here at the end of our text because it is the first concept needed in the whole field of multivariate analysis. We want to encourage you to go on to study this relatively new field because we think that it is a most important methodology for behavioral scientists. All we have been able to do here is help to get you started. Surely you can't stop now!

14.7 Summary

When the slope of the regression line is computed from standard score distributions of x and y, it turns out that

$$b_{z_y \cdot z_x} = b_{z_x \cdot z_y} = r_{xy} = \frac{\sum z_y z_x}{N}$$

The resulting correlation coefficient, r, is the average cross product of the standard scores. It has the property that

$$0 \leq |r| \leq 1$$

The variance of the predicted scores, z_y' or z_x', is r^2, and the standard error of estimate is $\sqrt{1 - r^2}$.

A useful interpretation is that r^2 is the proportion of the unknown variance in one variable that can be predicted from the known variance in the other variable.

When the variables can be assumed to be jointly normally distributed, the density function for the bivariate normal distribution can be used to compute $g(x, y)$, the density of the joint events $x_i = x$ and $y_i = y$. All x and y combinations for which g is a constant trace an ellipse in the bivariate space.

The null hypothesis that the obtained sample r_{xy} was from a population where $\rho_{xy} = 0.00$ can be tested using the t distribution, since

$$t = \sqrt{\frac{r^2(N-2)}{1-r^2}}$$

for normally distributed X and Y.

EXERCISES AND QUESTIONS

1. The correlation between variables X and Y is .55. The correlation between X and Z is $-.72$. Would you get better predictions of X using Y as a predictor or Z as a predictor? Why?

2. How would you expect age to correlate with reading comprehension ability for grade 12 students: positively, negatively, or no significant relationship? If it is negative, what does this imply? Using List 14.2, test your hypothesis about the nature of this relationship.

3. A scientist has just completed a study of 62 people who were measured on 15 variables. His main research question was concerned with the possible relationships among those 15 variables. He computed the 105 different correlations between all possible pairs ($15.14/2 = 105$) among the 15 variables. He found that the only correlation that seemed large enough to report was a correlation of .32 between 2 of the variables. All the others were between $-.2$ and $+.2$. Should he report the largest correlation and ignore the others?

4. LIST 14.1 is set up to distribute empirical t values for a two-tail test of the null hypothesis, as prescribed by the equation for t in the text. Modify the program to give the t distribution for a one-tail null hypothesis, as provided by the equation

$$t = r\sqrt{\frac{N-2}{1-r^2}}$$

SUMMARY

```
C      LIST 14.1   MONTE CARLO ON CORRELATION
C
C      THIS PROGRAM COMPUTES T WITH 8 DEGREES OF FREEDOM FOR THE NULL
C      HYPOTHESIS THAT R EQUALS ZERO, FOR 200 RANDOM SAMPLES OF SIZE 10
C      THE SINGLE DATA CARD PROVIDES A STARTING NUMBER FOR RANDOM IN
C      COLUMNS 1-8.
C      REQUIRED SUBROUTINES ARE RANDOM, RANDEV, UNIDIS, AND GOOFIT.
C
       DIMENSION R(1000), T(1000), F(100), IC(100), IE(100)
C
1      N = 10
       NDF = 8
       EN = N
       ENDF = NDF
       WRITE (6,2)
2      FORMAT (53H1 MONTE CARLO ON CORRELATION   WITH NDF FOR T = 8
       Y = RANDOM (1)
C
       DO 5   K = 1, 200
       S1 = 0.0
       S2 = 0.0
       SS1 = 0.0
       SS2 = 0.0
       SS12 = 0.0
       DO 4   J = 1, N
       X1 = RANDEV (0)
       X2 = RANDEV (0)
       S1 = S1 + X1
       S2 = S2 + X2
       SS1 = SS1 + X1 * X1
       SS2 = SS2 + X2 * X2
4      SS12 = SS12 + X1 * X2
C
       R(K) = (SS12-S1*S2/EN)/ SQRT((SS1-S1*S1/EN)*(SS2-S2*S2/EN))
5      T(K) = SQRT((R(K)*R(K)*ENDF)/(1.0-R(K)*R(K)))
C
       N = 200
       M = 10
       F(1) = -.00
       U = .20
       DO 3   J = 2, M
3      F(J) = F(J-1) + U

       WRITE (6,6)
6      FORMAT (50H0 DISTRIBUTION OF CORRELATIONS                    )
       CALL UNIDIS (N, M, R, F, IC)
C
       F(1) = .000
       F(2) = .141
       F(3) = .262
       F(4) = .399
       F(5) = .546
       F(6) = .706
       F(7) = .889
       F(8) = 1.108
       F(9) = 1.397
       F(10) = 1.860
       INC = 20
       DO 7   J = 1, M
7      IE(J) = INC
       WRITE (6,8)
8      FORMAT (50H0 DISTRIBUTION AND FIT FOR T TEST ON CORRELATIONS )
       CALL UNIDIS (N, M, T, F, IC)
       CALL GOOFIT (M, F, IE, IC)
C
       CALL EXIT
       END
```

```
C        LIST 14.2      CORRELATION AND REGRESSION ANALYSIS
C
C           THIS PROGRAM READS A PAIR OF SCORES FOR EACH OF N SUBJECTS AND
C        COMPUTES THE CORRELATION COEFFICIENT, THE T TEST FOR THE NULL HYPOTHESIS THAT
C        RHO EQUALS ZERO, AND ANALYZES THE REGRESSION OF EACH VARIABLE ON THE OTHER IN
C        TERMS OF STANDARD SCORE AND RAW SCORE REGRESSION EQUATIONS AND STANDARD
C        ERRORS OF ESTIMATE.
C
C           THE FIRST INPUT CARD SPECIFIES THE RUN NUMBER (COLS 1-5), N (COLS (6-10)
C        THE NAME OF THE FIRST VARIABLE (COLS 11-16), AND THE NAME OF THE SECOND
C        VARIABLE (COLS 17-22).  THE SECOND CARD GIVES THE FORMAT FOR THE SCORE CARDS,
C        BEGINNING WITH THE LEFT PARENTHESIS IN COLUMN ONE, FOR EXAMPLE, THIS CARD
C            (10X, F5.0, 5X, F5.0)
C        (BE SURE TO START IN COLUMN ONE) SPECIFIES THAT THE FIRST SCORE IS IN
C        COLUMNS 11-15 AND THE SECOND SCORE IS IN COLUMNS 21-25.  THE SCORES ARE
C        PRESENTED ON ONE CARD PER SUBJECT.
C
         DIMENSION   FMT(12)
C
         READ (5,2)   NPROB,  N,   XNAME,   YNAME
    2    FORMAT (2I5,  2A6)
         WRITE(6,3)    NPROB,   N,    XNAME,   YNAME
         WRITE(7,3)    NPROB,   N,    XNAME,   YNAME
    3    FORMAT(8H1PROBNO.I6,2H, I6,21H SUBJECTS, VARIABLES A6,5H AND A6)
         READ (5,4)   (FMT(J),    J = 1, 12)
    4    FORMAT (12A6)
         EN = N
         NDF = N - 2
         ENDF = NDF
         SX = 0.0
         SY = 0.0
         SSX = 0.0
         SSY = 0.0
         SSXY = 0.0
C
         DO 5   J = 1, N
C
         READ (5,FMT)  X,  Y
C    FMT(J)  CONTAINS THE VARIABLE INPUT FORMAT FOR SCORES.
C
         SX = SX + X
         SY = SY + Y
         SSX = SSX + X * X
         SSY = SSY + Y * Y
    5    SSXY = SSXY + X * Y
C
```

SUMMARY

```
      XM = SX / EN
      YM = SY / EN
      RNUM = EN * SSXY  -  SX * SY
      DSSX = EN * SSX   -  SX * SX
      DSSY = EN * SSY   -  SY * SY
      VX = DSSX / (EN * (EN - 1.0))
      VY = DSSY / (EN * (EN - 1.0))
      SDX = SQRT  (VX)
      SDY = SQRT  (VY)
      R = RNUM / SQRT  (DSSX * DSSY)
      RSQ = R * R
      T = SQRT  ((RSQ * ENDF) / (1.0 - RSQ))
C
      SEZ = SQRT  (1.0 - RSQ)
      SEXY = SDX * SEZ
      SEYX = SDY * SEZ
      AX = RNUM / DSSY
      CX = (SSY * SX  -  SY * SSXY) / DSSY
      AY = RNUM / DSSX
      CY = (SSX * SY  -  SX * SSXY) / DSSX
C
      WRITE(6,6)    XNAME,   XM,   VX,   SDX
      WRITE(7,6)    XNAME,   XM,   VX,   SDX
      WRITE(6,6)    YNAME,   YM,   VY,   SDY
      WRITE(7,6)    YNAME,   YM,   VY,   SDY
    6 FORMAT(10H0VARIABLE A6,7H MEAN =F8.3,12H, VARIANCE =F8.3,8H, S.D.
     C=F8.3)
      WRITE(6,7)    R,   RSQ,   SEZ
      WRITE(7,7)    R,   RSQ,   SEZ
    7 FORMAT(4H0R =F7.4,12H, R SQUARE =F7.4,27H,  Z-SCORE S. E. ESTIMATE
     C =F7.4)
      WRITE(6,8)    NDF,  T
      WRITE(7,8)    NDF,  T
    8 FORMAT(14H0T FOR R, WITHI6,5H DF =F7.3)
C
      WRITE(6,9)    XNAME,   YNAME
      WRITE(7,9)    XNAME,   YNAME
    9 FORMAT(29H0FOR RAW-SCORE REGRESSION OF A6,4H ON A6)
      WRITE(6,10)    AX,  CX
      WRITE(7,10)    AX,  CX
   10 FORMAT(16H0    A (SLOPE) = F8.3,19H,   C (INTERCEPT) = F8.3)
      WRITE(6,11)    SEXY
      WRITE(7,11)    SEXY
   11 FORMAT(14H0    SE X.Y =F8.3)
C

      WRITE(6,9)    YNAME,   XNAME
      WRITE(7,9)    YNAME,   XNAME
      WRITE(6,10)   AY,  CY
      WRITE(7,10)   AY,  CY
      WRITE(6,12)   SEYX
      WRITE(7,12)   SEYX
   12 FORMAT(14H0    SE Y.X =F8.3)
C
      CALL EXIT
      END
```

CHAPTER FIFTEEN

An overview of statistics

15.1 Inferring and estimating relationships

"A statistic is a value calculated from an observed sample with a view to characterizing the population from which it is drawn" (Fisher, 1950, p. 41). In this pithy sentence, one of the greatest of all statisticians defined our problem for us. To understand the sentence requires denotations for the technical terms "sample" and "population," and some connotations for "characterizing." These have been provided, but the student should realize that for as long as he continues to be interested in the study of statistics he will continue to add to his insights into the meaning of Fisher's definition. We hope you will sustain an interest in the expanding universe of meanings for the phrase "characterizing the population" for the rest of your life.

We have attempted in the writing of this brief course to convey to you not only the basic elements of this universe but also some of the excitement, mystery, and power inherent in this intellectual creation. From the many applications of statistics which you find in your reading of literature of the behavioral sciences, and even from the examples of actual research described in this text, you can induce the practical importance of statistical science, viewed as a methodology. Gradually and through much puzzling thought you will also develop an appreciation of the importance of statistics as a theory of knowledge. The statistical way of thinking about statements that claim to describe reality tends to develop intellectual virtues such as tolerant acceptance of incompleteness and ambiguity of knowledge, respect for both evidence and logic, adherence to the goals of predictivity and parsimony, and good-humored courage in the face of the essential uncertainty of almost everything. If statistical theory encourages these virtues it is worth studying for its own sake, as a palliative to the influence of other epistemologies which saturate human experience and nurture the spiritually gross and socially dangerous vices of dogmatism, bigotry, and certitude.

The "observed sample" of which Sir Ronald Fisher speaks is a part of the membership of a total set of elements of which we desire knowledge. This total set of elements is the "population." For one reason or another it is not feasible to study all the elements of the population. Either there are too many of them, or the process of studying an element leads to its destruction or to damaging it, or it is slow and costly to study each element. Thus, the first assumption of statistics is that we *must* be satisfied with incomplete and uncertain knowledge of the characteristics of the population we are interested in, derived from a study of a sample of elements selected from the population, because such a sampling study is either the only possible or the only reasonable type of study to do. When the scientist invokes statistical method he is good-humoredly renouncing omniscience as the object of his game. His good humor stems from his optimism regarding the usefulness of partial knowledge of a population which is the object of study. He recognizes that actions based on sound assessments of probabilities accumulate to increasing understanding of the population studied, and that in the long run the unavailable certainties are not much needed. Instead of seeking certainty, the scientist employing statistical method seeks to be as precise as possible regarding the probability that the statement he makes about the population is true.

Fisher speaks of a statistic as a value calculated from an observed sample. By value he means number. One of the restrictions on statistical methodology is that it requires quantification of the empirical data collected on the sample. The physical scientist is quite comfortable with this requirement that observations must be reduced to enumeration within categories (counting) or to mensuration along a continuum (scaling), but not all scholars concerned with the behavioral sciences are quite as comfortable. We are encouraged that the younger behavioral scientists as a group are more favorably inclined toward quantification than are their elders. Chapter 4 communicated our ideas on how the new availability of the digital computer as an auxiliary brain to scientists is accelerating the plunge into quantitative methods.

The fact that a statistic is a precise numerical summary of some aspect of the sample data should not conceal the inherent imprecision of the statistic in its more important role as an estimate of the corresponding characteristic of the population from which the sample was drawn. The most popular statistic is the sample mean. Sample means are often reported to several decimal places in a rather dazzling display of arithmetic precision. The point of Fisher's definition is that the sample mean is an estimate of the population mean, and its uncertainty or error in that role is not a function of the number of significant digits to which it is reported. It is a function of the sample size and of the variability on the measured characteristic of the elements of the population, and of the care with which the sample members are selected from the population. All these considerations have been discussed in Chapter 8, but the last one deserves further attention now, for it takes its place with *uncertainty* and *quantification* as another basic condition of statistical method. When Fisher speaks of the sample as drawn from the population he does not have in mind the act of taking for study the most available members of the population. College professors who have to try to do behavioral science research on a fiscal shoestring too often take an available, captive class of college students as subjects, and call this sample a representative sample of the human race, or of the subspecies "college students in the United States." Fisher's term "drawn" is elliptical for "drawn at random," and refers to an unbiased sample selected at random from the population. Basically this principle of *randomization* means that every element of the population has an opportunity to appear in the sample drawn, so that chance and chance alone determines the membership in the sample. The requirement of randomization can never be perfectly satisfied in the design of important behavioral research, but it remains an essential assumption of statistical reasoning, and it behooves the scientist to achieve as good an approximation to it in the selection of his subjects as he possibly can.

There are several ways in which a statistic derived from a sample may characterize the population which has been sampled. We have already mentioned the most obvious kind of fact-finding which a statistic may accomplish, in that it may provide an estimate of some desired quantitative generalization of some aspect of the distribution of frequencies or measurements in the population. Thus the population mean may be estimated by the sample mean. Another quantitative generalization about the distribution of values of a variable in the population which may be estimated from the sample data is an index of variability, or scatter, of the values around the mean, the "variance." Also estimable from a sample is the most useful of all numerical generalizations, Karl Pearson's "correlation coefficient," which we have seen is an index of the strength of relationship between two variables representing different but not necessarily independent characteristics of the population. Much of the burden of this first course in statistics has been to teach you how the mean, variance, and correlation coefficient describe the distributions of variables and the relationships among variables. Such numerical summaries or descriptions of characteristics of population distributions are termed "parameters," and getting the best possible estimates of population parameters from sample data is one of the tasks of statistics.

Statistics which qualify as "best estimates" of parameters have certain qualities. They are *consistent*, signifying that if a statistic is computed for many, many different random samples of a population, the mean of the resulting sampling distribution of the statistic will occur at the population value of the parameter being estimated. Actually, for any finite number of samples, the mean of a consistent statistic only tends to occur at the population value of the parameter, but as the number of samples over which the statistic is averaged increases this tendency becomes stronger. Another way of saying that a statistic is consistent is to say that it is an unbiased estimate, which is to say that it does not systematically overestimate or underestimate the parameter.

Best estimates are also *efficient*, signifying that as sample size increases the average squared error of estimation decreases, and that for any particular sample size the average squared error of estimation is smaller than it would be for any other statistic which might be proposed as an alternative estimator of the parameter under study. The concept of the average squared error of estimation may be arrived at by considering that each statistic computed for a random sample differs from the parameter by a random amount, which is its error. Some of these errors are positive, indicating an overestimate, and some are negative, indicating an underestimate, so that in the sampling distribution of the statistic the sum of errors, and thus the average error, tends to zero. For this reason our concept of the efficiency of estimation involves the requirement that the average squared error be minimized, since all squares are positive, and there is no cancelling of positives by negatives. This concept of minimizing the average squared error also emphasizes the avoidance of larger errors, since larger numbers have much larger squares. This criterion of efficiency is known as the "least squares" criterion. We have put the computer to work generating sampling distributions of some statistics so that you could observe their means and average squared errors, or "error variances." We hope you will habitually inquire into the standard errors of statistics you encounter.

Yet another quality which best estimators usually possess is *sufficiency*, which indicates that they make use of all the available information in the sample data in estimating parameters. The descriptive statistics emphasized in this text are members of a class of statistics which R. A. Fisher has named "maximum likelihood statistics," and they satisfy the criteria we have been discussing. Statistics in this class can be shown by the differential calculus to provide estimates of the parameters of the particular populations from which the sample data have the highest likelihood, or probability, of arising, out of all the populations with different parameters from which the samples might have been drawn. This means that when we use these statistics we get estimates of the most likely population parameters, given our data. Sometimes the likelihood, or probability, that the parameter is what the statistic estimates it to be is disturbingly slight, but it remains the best estimate available from the sample drawn. The method of maximum likelihood and other mathematical methods of deriving and proving the properties of statistics are way beyond the scope of this text, but the student should be aware that there exists a vast literature of mathematical statistics containing the theoretical bases of the applications of statistical methods to research problems he has begun to study. If you are mathematically inclined, or if you want to try to become more mathematical in your approach to statistics, we recommend you follow up our presentation of elementary notions and procedures with self-study of the first eleven chapters of Hays' excellent text (Hays, 1963). Or if you have a considerable background in college mathematics, we recommend Mood and Graybill (1963).

The general approach to characterizing a population by a sample statistic may be called *point estimation* because it involves the calculation of the best single-valued estimate of a population fact. It has been hinted that sometimes this best estimate of the point along a continuum at which the parameter occurs may be exceedingly untrustworthy. Fortunately, the statistician pays the piper by providing an estimate of the standard error of estimation along with every statistic he delivers. By suitable manipulation of this error estimate and invoking a theory about the distribution of errors, the statistician is able to substitute for his point estimate an interval on the continuum within which the parameter has a specified probability of existing. Naturally, in any particular analysis wider intervals give higher probabilities. Of course, if the average squared error of estimation for the statistic is very large, a confidence interval of a reasonably large probability (say .95) may turn out to embrace most of the range of the possible values on the continuum, making it rather useless, but then perhaps the research was not well designed. To contrast the two approaches, point estimation produces the best single value for the parameter but one which may have a very low probability of being correct, while *interval estimation* yields a range of values within which the parameter has a specified probability of occurring. The former may be thought of as a precise estimate of unspecified trustworthiness, while the latter gives an imprecise estimate of specified trustworthiness. Both approaches have their proper applications, but it seems to the authors that behavioral scientists have been guilty of some degree of unsobriety in their pronounced preference for point over interval estimation. Interval estimation deserves to be used more because it makes a much more honest admission of the imprecision of the information gained from most behavioral research.

Yet another and rather different approach to characterizing the population by statistics based on a sample involves the testing of hypotheses about population parameters which the scientist formulates prior to the collection of the sample data. Such hypotheses are deduced from theory and available knowledge of the situations under study. Often they specify contrasts between corresponding parameters of two populations. An example which is suggested by available educational knowledge and supported by a theory of sex linkage in developed abilities is the hypothesis that high school girls tend to be better spellers than high school boys. A second hypothesis which suggests itself is that high school boys are more dispersed around their mean in spelling than are high school girls. Technically this second hypothesis states that the variance is larger for the boys than for the girls. Given these two hypothesized contrasts between corresponding parameters of spelling test scores for the two populations of high school girls and boys, the researcher does the best he can to draw a reasonably random sample of appropriate size from each population, and prepares a reasonably definitive spelling test for administration to his two samples. You can see that a host of research design problems are implied by the last sentence. Probably the researcher has to define his populations in terms of a limited geographic area, such as "high school youth in Massachusetts." Sample size is a complex function of the smallest differences in population parameters the researcher wishes to detect and the expected variability of the spelling scores. The precise definition of the concept "spelling" and its translation into an adequate operational variable through test construction and development techniques will require considerable expertise and resources. At last the data are collected and the sample means and variances are calculated. Perhaps it turns out that indeed the girls' mean is larger than the boys', and the boys' variance is larger than the girls'. But these are comparisons of sample statistics which are subject to error as estimates of population facts, and the researcher is not sure whether the contrasts between the sample statistics are indicative of similar contrasts between parameters or whether they can be explained as due to the operation of chance. Fortunately, statisticians are able to answer such questions by the application of procedures which come under the rubric of *statistical inference*, or hypothesis testing. The statistician will compute the probability that the sample contrasts arose by chance, on the assumption that the underlying parameters are equal in the two populations. If these probabilities are low enough the researcher will dare to assert to the scientific community that chance is not enough of an explanation of his sample statistics, and that true differences in the hypothesized directions seem to have been verified. It might happen that one of the research hypotheses would be substantiated and not the other. Perhaps the results would verify that the girls are on the average better spellers but would fail to verify that the boys are more variable in spelling than the girls.

Theory building is a very important aspect of science, and the testing of hypotheses deduced from theoretical formulations is the key to honest theorizing. The role of statistics in support of theory building in the behavioral sciences is analogous to that of the guidance system in a ballistic missile. The guidance system provides feedback to the missile about discrepancies between its intended course and its actual track, and enables the propulsion system to correct the track. Under control of its guidance system the missile streaks toward its proper target, reflecting in its passage the brilliant light of human intelligence acting at a distance. When the guidance system fails or goes wild the missile becomes an extraterrestrial, supersonic berserker threatening a visitation of its lethality on some randomly selected neighborhood. In its case human intelligence continues to act at a distance through the method of a destruct signal. If the analogy is disquieting, it is intended to be. Behavioral science properly monitored by the feedback mechanism of statistical inference portends the accrual of enormous control over the destiny of human intelligence. A berserk scientism, uncontrolled by testing mechanisms, hypocritically pilfering the prestige and respect accorded to true science but devoid of its integrity and modesty, can tyrannize humanity by establishing erroneous and destructive dogmas as "scientific truths." The genetic theory popularized by Nazi "science" comes to mind for its enormous destructive impact. The genetic theory imposed by Stalin comes to mind for its obtuse obstructionism. Who can say what enormities might have been avoided if the theories of Marx and Freud had been subject to greater degrees of testing in the courses of their development? A major purpose of this book has been to help you develop an appreciation of how statistical inference is applied in the testing of social and behavioral theories to keep them honest. Informed opinion can send a destruct signal to wild theory.

Statistical methods provide tools for the study of relationships among variables. Sometimes the relationship under study is between a classification variable, or group membership variable, on the one hand, and a personality trait continuously distributed among individuals on the other. Our example of an inquiry into sex (group) differences in spelling ability (continuous personality variable) is an example of the linkage between group membership and individual differences on a personality trait. Very often in psychology the relationship under study is between two personality variables or between a personality trait and a continuously distributed environmental trait. Examples of the former are studies of the relationships between intelligence on the one hand and school achievement traits and vocational interests on the other. Examples of the latter are studies of the relationship of intelligence to such environmental measures as family income and number of years of schooling completed by parents. Statistics answers two questions about such relationships: (1) does a nonzero relation exist between the variables, and (2) if it does, how strong and in what direction is it? The statistical approach to the first question is again an example of *inference*, and to the second question is another example of *estimation*. This is a worthwhile distinction if it is not carried too far. In fact, a good research design provides for both processes in the analysis of data. Inference answers the question of whether relationships observed in the sample data signify similar relationships in the population sampled. The alternative hypothesis is that the observed relationships have been established in the sample data by the operation of chance

alone through the mysterious workings of randomization. An observed relation which cannot reasonably be attributed to chance is said to be statistically significant, or to lead to the rejection of the *null hypothesis* of no similar relation in the population. Unfortunately, a statistically significant relation need not be intellectually significant as well. Many real relationships among variables in target populations may be so weak that they should be ignored or deemed essentially nonexistent in theory. As sample sizes increase, weaker and weaker relationships can be discovered as statistically significant relations in the sample data. Hence statistical inference by itself can be as misleading as it is helpful to the scientist.

The problem of the behavioral scientist is particularly acute because in an organism all the parts and processes interact in a highly organized, coherent fashion to support a meaningful whole we call a "life." In such a system every pair of attributes of an organism (person) is probably related to some extent, however slight. To catalog an endless list of very weak but statistically significant relationships among human attributes does not seem to be a fruitful undertaking. The authors are appalled at the willingness of many researchers to take the bald assertion that a relationship is statistically non-zero as sufficient corroboration for a theoretical deduction, without knowing the estimated strength of the relationship. Also, as the next section amplifies, the randomization assumed by statistical inference is often entirely missing from published research, and in these cases the act of statistical inference is superfluous. Statistics can be a confidence game in the worse sense, as well as a rigorous logic for estalishing degrees of confidence in statements. The authors urge you to insist that the statistical inference game be played approximately according to its rules when you are a participant, even a passive one out in the audience.

15.2 Random sampling and random assignment

The theory of errors of estimate on which statistics is built requires the assumption of a random distribution of errors. This assumption can only be justified when a randomizing operation has been conducted by the researcher. It is incumbent on the researcher to report in detail his randomizing procedure, so that the reader may judge its adequacy for himself. The fact that few research designs in the social sciences are able to achieve rigorous randomization underscores the need for honest publicity for the compromises which are deemed necessary. Many of these compromises could be avoided if adequate funding of research existed. Publicity may lead to increased awareness of the problem and greater support for improvement of the rigor of designs. Publicity might also force sloppy or lazy researchers to tighten up their designs and to be less ambitious and more rigorous in the deployment of their resources. When you read research reports you should pay close attention to the sections titled "Sampling Procedure" or "Research Design." You cannot know what population is characterized by the sample statistics until you are clear about the randomization scheme employed in the data collection. Remember that the purpose of the statistics in the research report is to encourage you to generalize from events in the sample to supposed events in the population. You want to acquire intelligent confidence in some generalizations which deserve your endorsement. You do not want to be "conned" into blind endorsement of unjustified generalizations by a shoddy display of statistical camouflage. The extent of your need for wariness depends in part on the corner of social science literature which you read, but we think you will have to be alert in any field. Fortunately, this alertness will bring the reward of high pleasure in the discovery of examples of well-designed researches incorporating reasonable approximations to randomization.

Two types of randomization appear in behavioral research. The first is *random sampling* of a target population. This has been accomplished when every member of the population had a greater than zero chance to appear in the sample. The obvious way to draw such a sample is to assign a serial number to every member of the population, starting with *1* and going to *N*, where *N* is the head count of all members, and then to use a random number generator to select *n* random integers in the range *1* through *N*, where *n* is the size of the desired sample. Each of the random integers is the identification number of a selected member of the sample. Random number generation on the computer has been described in Chapter 5. Sometimes this procedure is practical (for example, to get a random sample of 50 third graders from the population of 4383 third graders in city school system XYZ in October, 1965), but notice that it requires a listing of all the members of the population. Usually important populations are too numerous to be listed, or they are theoretically infinite, as when they include deceased and unborn members of a species. In the former case, it may be that good census data exist on the geographic and institutional distribution of the population, so that the areas or institutional units in which the population is dispersed can be randomly sampled, and a representative stratified random sample can be assembled. Extensive theory and technique are available to scientists confronted with this case. You can find an excellent nontechnical summary of area sampling methods in Travers (1964, Chapter 10), and you might look at Cochran (1963) to see what a full treatment of the subject looks like. Chapter 1 of this text described one example of a stratified random sample, the Project TALENT sample. The latter case of an assumed infinite population is appropriate only in a class of researches called "experiments" such as the HOSC study which was also discussed in Chapter 1.

The type of research which attempts a careful approximation to a random or probability sampling of a large population is the survey study. A *survey* tries to establish accurate estimates of the facts about a population. The U.S. Census Bureau does most of its fact-finding by sampling surveys, because complete census on most of the thousands of questions of interest to the Federal government and its clients would be ridiculously expensive and is unnecessary in the light of the high quality of the information returned from sampling surveys. The Kinsey studies of sexual behaviors in Americans are well-known examples of behavioral science surveys, although the sampling procedures employed have been criticized. An outstanding achievement of the method is Stouffer's study of American public opinion on citizenship issues, *Communism, Conformity, and Civil Liberties* (1955).

Surveys test hypotheses by observing the outcome of natural experiences of sample members. In doing a survey the scientist seeks to remain in the shadows and not disturb the natural life processes of his subjects in any excessive way. Another class of researches which are designated experiments involves the scientist in direct manipulation of his subjects. In an experiment, subjects are placed in different groups and each group is given a separate treatment. For example, in testing a drug the "experimental" group receives the drug while the "control" group receives the placebo. All other conditions are equated for the two groups. The questions for statistics are whether the group differences on the criterion variable are large enough to be attributed to the drug rather than to sampling error, and if so, what the estimated magnitude of the impact of the drug on the criterion variable is. The first question can only be answered when the second type of randomization, namely, *random assignment* of subjects to treatment groups, has been carried out. The second question is once again dependent on the random sampling of the population to which generalizations are to be made. Thus, an experiment involves both types of randomization operations. First, a pool of subjects is drawn by random sampling of a target population, and second, the members of the pool are randomly assigned to the two or more treatment groups. We have conceded that the first type of randomization can usually only be approximated, but there is seldom any excuse for failure to adhere rigorously to the second type of operation. Statistical inference can be applied to the outcome of experiments only when there has been strict randomization of assignment to treatment groups. This is one rule behavioral scientists can and should abide by. It makes the game they play recognizable. Chapter 1 described an example of an educational experiment in which classes of science students were randomly assigned to teaching methods and class means were the units of analysis.

You have many exciting adventures in the study of statistics before you, if you will persevere. The many subtleties which can be introduced into the design and analysis of experiments are treated in a subject called "the analysis of variance." In this the greatest of Fisher's inventions, it is possible to partition variance in outcomes of an experiment into many independent components assigned to different influences or sources. The complexities of data collection and analysis in advanced survey designs are treated in the subject of "multivariate analysis," wherein the simple bivariate correlation coefficient becomes the building block for the expression and testing of elaborate, intricate theories regarding the interrelationships among numerous variables. In both these fields, the computer takes the chore out of analysis and leaves a delightful and powerful game for the human intellect. We wish you pleasure and success in your continued exploration of the role of statistics in the science of human behavior.

References

Abramowitz, M., and Stegun, I. A. (Editors). *Handbook of mathematical functions with formulas, graphs, and mathematical tables.* Washington: U.S. Government Printing Office, 1964.
Bauer, R. A. Note. *Amer. Psychologist*, 1964, **19**, 288–289.
Bloom, B. S. *Stability and change in human characteristics.* New York: John Wiley and Sons, 1964.
Cochran, W. G. *Sampling techniques* (2nd ed.). New York: John Wiley and Sons, 1963.
Cohen, J. The statistical power of abnormal-social psychological research. A review. *J. of Abnorm. Soc. Psychol.*, 1962, **65**, 145–154.
Cooley, W. W., and Jones, K. J. Note. *Amer. Psychologist*, 1965, **20**, 298.
Cooley, W. W., and Klopfer, L. E. *Test on understanding science.* Princeton: Educational Testing Service, 1961.
Cooley, W. W., and Klopfer, L. E. The evaluation of specific educational innovations. *J. Res. Sci. Tchg.*, 1963, **1**, 73–80.
Cooley, W. W., and Lohnes, P. R. *Multivariate procedures for the behavioral sciences.* New York: John Wiley and Sons, 1962.
Edwards, A. L. Social desirability and performance on the MMPI. *Psychometrika*, 1964, **29**, 295–308.
Feigenbaum, E. A., and Feldman, J. (Editors). *Computers and thought.* New York: McGraw-Hill, 1963.
Fisher, R. A. *The design of experiments* (6th ed.). New York: Hafner, 1950.
Flanagan, J. C., et al. *Design for a study of American youth.* Boston: Houghton Mifflin, 1962a.

REFERENCES

Flanagan, J. C., et al. *Studies of the American high school.* Pittsburgh: Project TALENT Office, 1962b.

Flanagan, J. C., et al. *The American high school student.* Pittsburgh: Project TALENT Office, 1964.

Flanagan, J. C., et al. *Project TALENT one-year follow-up studies.* Pittsburgh: Project TALENT Office, 1966.

Galler, B. A. *The language of computers.* New York: McGraw-Hill, 1962.

Green, B. F., Jr. *Digital computers in research: An introduction for behavioral and social scientists.* New York: McGraw-Hill, 1963.

Gribbons, W. D., and Lohnes, P. R. Relationships among measures of readiness for vocational planning. *J. Counsel. Psych.*, 1964, **11**, 13–19.

Hays, W. L. *Statistics for psychologists.* New York: Holt, Rinehart and Winston, 1963.

Huff, D. *How to lie with statistics.* New York: Norton, 1954.

IBM. *Reference manual: Random number generation and testing.* C 20–8011, 1959.

Kempthorne, O. The randomization theory of experimental inference. *J. Amer. Statist. Assoc.*, 1955, **50**, 946–967.

Kerlinger, F. N. *Foundations of behavioral research.* New York: Holt, Rinehart and Winston, 1964.

Klopfer, L. E., and Cooley, W. W. Use of case histories in the development of student understanding of science and scientists. Cooperative Research Project Number 896, U.S. Office of Education, 1961 (mineographed).

Klopfer, L. E., and Cooley, W. W. The history of science cases for high schools in the development of student understanding of science and scientists. *J. Res. Sci. Tchg.*, 1963, **1**, 33–47.

Lohnes, P. R., and McIntire, P. H. Classification validities of a statewide 10th grade test program. *Pers. Guid. J.* 1967, 561-567.

Lohnes, P. R. *Measuring Adolescent Personality.* Pittsburgh: Project TALENT, 1966.

Mood, A. M., and Graybill, F. A. *Introduction to the theory of statistics* (2nd ed.). New York: McGraw-Hill, 1963.

Parzen, E. *Modern probability theory and its applications.* New York: John Wiley and Sons, 1960.

Shaycoft, Marion F., et al. *Studies of a complete age group—age 15.* Pittsburgh: Project TALENT Office, 1963.

Siegel, S. *Nonparametric statistics.* New York: McGraw-Hill, 1956.

"Student," The probable error of a mean. *Biometrika*, 1908, **6**, 1–25.

Terman, L. *Genetic studies of genius.* Stanford, Calif.: Stanford University Press, 1947.

Travers, R. M. W. *An introduction to educational research* (2nd ed.). New York: Macmillan, 1964.

Wert, J. E., Neidt, C. O., and Ahmann, J. C. *Statistical methods in educational and psychological research.* New York: Appleton-Century-Crofts, 1954.

APPENDIX A

Symbol conventions

X_i	Score on variable X for individual i.
N	Total number of observations or trials (sample size).
N_j	Number of observations in category j.
P_j	Proportion of cases in category j.
M_x	Sample mean of variable X.
μ_x	Population mean of variable X.
Σ	Signifies the operation of addition.
x_i	Deviation from mean for individual i.
σ_x^2	Population variance of X.
s_x^2	Variance for the sample in hand (that is, sample variance).
$\hat{s}_x^2$	Unbiased estimate of the population variance of X.
σ_x	Standard deviation of X for the population.
s_x	Sample standard deviation.
$\hat{s}_x$	Unbiased estimate of population standard deviation of X.
z	Standard score with mean of zero and standard deviation of 1.
T	Standard score with mean of 50 and standard deviation of 10.
$p(X)$	Probability of obtaining a particular score X (used only with discrete variables or intervals of a continuous variable).
$P(X)$	Cumulative probability of obtaining a score equal to or less than X.
$N!$	N factorial, meaning the continuous product from 1 to N $(1 \cdot 2 \cdot 3 \cdot \ldots (N-1) \cdot N)$.
$X \neq Y$	X is not equal to Y.
$X < Y$	X is less than Y.
$X \leq Y$	X is less than or equal to Y.

$X > Y$	X is greater than Y.
$X \cong Y$	X is approximately equal to Y.
e	Base of the natural logarithms (2.72).
Π	Pi, a constant, 3.14159265.
$\sigma_M{}^2$	Sampling variance of the mean for the population of means.
$s_M{}^2$	Estimate of $\sigma_m{}^2$.
σ_M	Standard error of the mean for the population of means.
s_M	Estimate of σ_m.
$\|x\|$	Absolute value of x, regardless of sign.
χ^2	Chi squared, a theoretical distribution.
t	Student's t, a theoretical distribution.
r_{xy}	Sample correlation between X and Y.
ρ_{xy}	Correlation between X and Y in population.
α	Probability of rejecting a null hypothesis when it is true.
β	Probability of accepting a null hypothesis when it is false.
$s^2_{x \cdot y}$	Variance error of estimate for regression of x on y.
$s_{x \cdot y}$	Standard error of estimate for regression of x on y.

APPENDIX B

Project TALENT data

Table B.1 Selected Project TALENT Variables

Variable Number	Card Columns	Name of Variable
ID	2–4	Student Identification Number
1	7	School Size (4 categories—based on number of seniors) 1. Under 25 2. 25–99 3. 100–399 4. 400 or more
2	9	Geographic Region (9 categories) 1. New England —6 States 2. Mid east —5 States and D.C. 3. Great Lakes —5 States 4. Plains —7 States 5. Southeast —12 States 6. Southwest —4 States 7. Rocky Mountains—5 States 8. Far West —4 States 9. Noncontiguous —Alaska and Hawaii
3	11–12	Age (nearest year)
4	14	Sex (1 = male, 2 = female)
5	16–17	Career Plans (36 vocations) 01. Accountant 02. Biological scientist (biologist, botanist, physiologist, zoologist, etc.) 03. College professor 04. Dentist 05. Engineer (aeronautical, civil, chemical, mechanical, etc.) 06. Elementary school teacher 07. High school teacher 08. Lawyer 09. Mathematician 10. Pharmacist 11. Clergyman (minister, priest, rabbi, etc.) 12. Physical scientist (chemist, geologist, physicist, astronomer, etc.) 13. Physician 14. Political scientist or economist 15. Social worker 16. Sociologist or psychologist 17. Armed forces officer

Table B.1 (cont.)

Variable Number	Card Columns	Name of Variable
		18. Artist or entertainer
		19. Businessman
		20. Craftsmad
		21. Engineering or scientific aide
		22. Forester
		23. Medical or dental technician
		24. Nurse
		25. Pilot, airplane
		26. Policeman or fireman
		27. Secretary, office clerk or typist
		28. Writer
		29. Barber or beautician
		30. Enlisted man in the armed forces
		31. Farmer
		32. Housewife
		33. Salesman or saleswoman
		34. Skilled worker (electrician, machinist, plumber, printer, etc.)
		35. Structural worker (bricklayer, carpenter, painter, paperhanger, etc.)
		36. Some other occupation different from any of the above
6	19–20	Weight (pounds)
		01. 74 or less
		02. 75–89
		03. 90–104
		04. 105–119
		05. 120–134
		06. 135–149
		07. 150–164
		08. 165–179
		09. 180–194
		10. 195–209
		11. 210–224
		12. 225 or more
7	22–23	Type of College Student Plans to Attend (9 categories)
		01. Do not expect to go to college
		02. Teachers college
		03. Agricultural college
		04. Engineering college
		05. Liberal Arts college
		06. College specializing in music or fine arts
		07. University which includes many of the above colleges
		08. Some other type of college
		09. Have no plans regarding the type of college I will attend
8	25	Plan College Full time? (SIB 301)
		1. Definitely will go
		2. Almost sure to go
		3. Likely to go
		4. Not likely to go
		5. Definitely will not go
9	27–29	Information Test, Part I (R 190)

Table B.1 (cont.)

Variable Number	Card columns	Name of Variable
10	31–33	Information Test, Part II (R-192)
11	35–37	English Test (R-230)
12	39–40	Reading Comprehension Test (R-250)
13	42–43	Creativity Test (R-260)
14	45–46	Mechanical Reasoning Test (R-270)
15	48–49	Abstract Reasoning Test (R-290)
16	51–52	Mathematics Test (R-340)
17	54–55	Sociability Inventory (R-601)
18	57–58	Physical Science Interest Inventory (P-701)
19	60–61	Office Work Interest Inventory (P-713)
20	63–65	Socioeconomic Status Index (P-801)

Table B.2 Project TALENT Data—20 Variables for 234 Twelfth-Grade Males

ID	1	2	3	4	5	6	7	8	9	10	11	12	13	14	15	16	17	18	19	20
1	3	1	17	1	22	05	01	4	151	078	087	39	09	12	09	20	10	20	18	103
2	2	1	18	1	34	06	01	5	100	041	076	15	07	10	10	15	04	15	13	088
3	3	1	17	1	19	07	08	5	156	069	090	28	08	12	09	26	09	08	06	105
4	1	1	18	1	36	05	08	3	164	089	082	47	13	14	12	29	04	28	24	091
5	3	1	17	1	04	06	07	1	164	088	094	40	10	15	12	32	11	26	01	097
6	4	1	17	1	18	06	01	5	118	071	086	21	10	14	11	21	06	08	09	099
7	3	1	17	1	36	08	08	1	123	072	076	33	09	12	09	25	11	16	11	113
8	2	1	17	1	12	07	05	2	224	096	099	46	18	20	15	51	09	36	02	098
9	3	1	17	1	34	05	09	4	162	071	093	42	10	17	13	31	06	33	16	095
10	2	1	17	1	05	09	04	2	183	082	079	38	14	18	11	39	09	30	03	102
11	3	1	17	1	05	07	07	1	196	084	085	42	12	17	12	32	06	27	12	111
12	2	1	17	1	07	09	07	2	180	073	082	32	10	18	08	31	01	20	23	106
13	3	1	18	1	34	11	01	5	216	103	087	39	16	17	11	34	04	23	11	092
14	3	1	18	1	07	07	02	1	182	095	081	43	08	10	11	34	08	28	19	092
15	4	2	17	1	12	11	05	5	191	095	099	41	13	10	08	34	07	33	04	099
16	4	2	17	1	34	06	02	5	123	072	076	34	07	09	05	16	08	09	06	101
17	4	2	17	1	07	06	02	4	185	083	091	41	11	12	11	32	02	17	20	094
18	4	2	17	1	04	06	05	2	165	073	092	38	11	14	11	35	07	18	14	106
19	4	2	17	1	07	08	02	3	147	074	081	32	05	14	13	30	04	13	03	102
20	4	2	17	1	36	07	07	1	192	096	092	41	17	17	11	27	08	21	20	102
21	4	2	16	1	36	07	09	4	139	065	086	32	10	12	07	15	08	25	03	092
22	4	2	17	1	07	09	02	1	193	081	086	43	05	11	11	42	02	28	14	090
23	3	2	17	1	05	08	01	4	116	070	067	24	09	09	07	16	05	27	24	092
24	4	2	17	1	19	06	07	1	187	107	092	43	12	15	12	37	11	26	06	109
25	4	2	16	1	05	06	04	1	208	091	092	43	16	19	12	39	02	31	14	117
26	4	2	17	1	36	10	01	3	125	061	078	25	10	15	07	23	08	17	14	089
27	4	2	16	1	34	07	08	4	190	093	083	36	14	16	12	39	03	24	16	096
28	4	2	16	1	05	07	04	1	196	092	097	45	10	16	11	49	09	34	17	100
29	4	2	17	1	07	07	07	1	131	076	076	27	08	10	13	17	11	24	12	100
30	3	2	17	1	02	07	07	1	210	089	099	39	09	17	11	44	09	28	10	112
31	3	2	17	1	13	05	05	2	207	100	096	44	18	15	10	43	08	27	11	120
32	3	2	17	1	36	08	01	4	167	083	083	36	13	11	08	10	03	24	19	105
33	2	2	18	1	19	06	05	3	167	070	085	35	07	15	11	27	03	23	23	103
34	2	2	18	1	30	04	01	4	145	072	082	27	10	12	11	19	08	18	11	085
35	3	2	17	1	04	07	07	1	210	093	089	43	16	17	10	42	12	34	14	113
36	3	2	17	1	12	07	07	1	190	091	104	47	08	13	14	47	10	34	13	114
37	3	2	18	1	21	08	04	3	172	081	084	36	18	16	08	18	09	26	03	095
38	2	3	17	1	08	06	05	4	165	096	088	42	13	12	14	28	09	18	10	084
39	3	5	17	1	04	08	07	1	194	096	085	41	15	19	12	41	05	24	19	114
40	3	2	17	1	22	07	07	1	197	096	086	44	14	20	12	37	05	19	11	107
41	3	2	18	1	34	05	01	5	149	079	090	43	09	17	13	32	05	14	06	101
42	3	2	18	1	34	08	05	1	170	095	094	40	13	15	06	23	07	34	09	096
43	3	5	17	1	10	07	05	1	214	095	099	44	17	20	10	32	11	30	23	095
44	4	2	16	1	18	06	06	3	107	062	071	23	01	05	09	15	07	06	04	089
45	4	2	16	1	36	06	08	3	141	069	089	33	07	16	11	24	07	18	16	095
46	4	2	17	1	18	08	06	2	200	106	106	48	18	13	12	37	05	11	00	094
47	4	2	18	1	36	06	01	5	137	068	071	34	07	15	11	14	06	06	04	087
48	4	2	17	1	36	06	01	4	095	053	079	23	06	08	08	09	06	19	23	086

ID	1	2	3	4	5	6	7	8	9	10	11	12	13	14	15	16	17	18	19	20
49	4	2	17	1	19	06	09	3	146	087	088	39	14	12	12	36	08	18	23	096
50	2	2	17	1	05	05	05	3	191	082	082	36	12	16	11	39	05	40	21	082
51	4	2	17	1	36	12	01	5	124	076	069	23	11	16	12	13	06	23	07	089
52	3	2	17	1	36	06	01	4	138	056	089	25	07	11	08	13	09	21	20	080
53	2	2	17	1	15	07	01	5	080	051	047	24	07	08	04	11	06	13	13	086
54	3	2	17	1	36	12	01	5	149	088	074	22	05	14	08	16	10	11	07	103
55	2	2	18	1	31	07	01	1	131	066	059	26	06	16	08	22	03	12	13	084
56	3	2	17	1	26	07	01	4	105	056	069	05	06	09	08	14	07	15	16	101
57	3	2	17	1	36	05	01	4	178	086	067	29	09	13	09	13	03	14	01	086
58	3	2	17	1	30	06	01	3	149	087	088	45	14	12	10	33	02	24	24	093
59	2	2	18	1	17	06	01	4	162	077	099	38	12	16	11	27	01	16	01	091
60	2	2	18	1	36	06	01	5	162	082	088	44	11	15	12	27	02	18	24	113
61	2	2	17	1	25	06	08	5	155	083	082	44	07	14	08	27	07	31	01	096
62	3	2	17	1	05	05	04	1	191	095	072	43	16	16	09	33	12	36	06	101
63	3	2	17	1	05	10	07	1	220	112	100	46	13	16	09	50	08	36	07	111
64	2	2	18	1	35	06	01	4	163	072	080	34	10	17	11	24	09	20	14	091
65	3	3	17	1	05	06	01	4	120	066	075	34	10	12	10	13	08	19	14	100
66	3	3	17	1	30	07	01	5	199	113	084	40	19	15	09	32	03	29	08	099
67	3	3	18	1	05	10	04	1	197	088	100	36	09	14	09	41	08	36	09	108
68	3	3	17	1	18	07	09	4	106	063	078	20	05	08	08	21	07	02	01	094
69	4	3	17	1	09	11	04	1	214	107	094	44	11	10	10	44	08	31	09	115
70	4	3	17	1	05	07	04	5	178	098	098	33	14	11	09	36	10	31	13	101
71	4	3	18	1	34	07	01	5	141	069	074	28	08	15	08	19	05	13	01	093
72	4	3	17	1	36	07	05	1	218	105	095	39	13	15	11	35	05	24	04	109
73	2	3	17	1	05	08	07	2	177	083	086	39	14	18	15	39	11	33	11	114
74	4	3	17	1	04	08	04	1	185	094	078	44	11	18	10	39	05	25	09	102
75	4	3	18	1	26	06	01	4	116	055	064	28	01	10	03	23	06	21	17	105
76	4	3	17	1	05	05	04	3	162	091	086	37	08	13	07	15	10	26	13	101
77	3	3	17	1	13	06	07	1	202	081	094	45	11	15	14	46	11	26	14	103
78	2	3	19	1	36	07	01	4	120	065	074	20	05	09	09	15	03	21	03	103
79	2	3	17	1	01	05	08	4	115	062	073	32	09	15	10	26	09	08	27	095
80	2	3	17	1	18	07	09	3	168	098	097	38	10	10	12	25	02	11	08	097
81	2	3	17	1	34	05	01	5	124	063	078	28	08	11	06	20	12	17	04	106
82	2	3	17	1	07	07	05	2	164	072	093	39	13	12	12	43	06	28	17	100
83	3	3	17	1	02	07	07	1	168	085	078	27	07	12	09	24	10	35	14	115
84	1	3	18	1	35	08	01	4	122	059	075	30	14	13	11	11	10	10	10	078
85	3	3	17	1	34	06	01	5	122	055	071	24	08	16	12	17	06	23	10	101
86	3	3	17	1	12	06	05	1	217	117	102	44	18	15	14	49	10	33	06	115
87	3	3	17	1	01	07	05	1	197	103	098	44	13	09	11	45	11	33	24	120
88	3	3	17	1	09	08	09	4	112	049	056	16	05	11	05	12	05	20	30	088
89	4	3	17	1	02	07	03	1	221	110	094	47	15	14	10	36	03	37	13	110
90	4	3	16	1	12	07	05	2	193	077	100	42	13	14	11	44	07	34	17	096
91	3	3	19	1	05	05	04	3	146	060	073	25	14	10	08	27	10	26	06	093
92	2	3	17	1	05	06	02	1	130	074	080	30	06	11	10	26	11	29	16	092
93	3	3	17	1	34	06	01	4	139	082	079	24	11	19	06	24	10	14	14	103
94	3	3	17	1	05	08	04	1	217	115	100	43	16	17	14	49	05	30	03	121
95	3	3	17	1	19	10	05	2	204	102	098	44	15	17	09	31	01	28	13	095
96	4	3	17	1	36	06	01	5	114	053	075	29	03	18	12	25	09	09	01	099

ID	1	2	3	4	5	6	7	8	9	10	11	12	13	14	15	16	17	18	19	20
97	4	3	18	1	33	07	09	3	162	089	083	42	10	17	11	17	04	07	04	102
98	4	3	17	1	19	06	05	1	140	075	087	41	14	16	13	40	04	13	24	098
99	2	3	17	1	06	06	02	5	155	071	082	28	12	15	07	27	09	34	26	100
100	3	3	18	1	12	06	07	2	170	074	075	21	11	14	10	12	08	34	16	099
101	3	3	17	1	30	05	01	4	142	086	081	27	12	16	07	19	08	21	13	099
102	2	3	17	1	12	08	01	5	183	079	092	35	11	13	11	27	01	16	00	099
103	2	3	18	1	01	02	01	5	147	087	078	31	07	12	02	16	12	08	07	113
104	4	3	17	1	13	08	05	1	173	084	076	39	14	18	10	36	07	24	17	114
105	4	3	17	1	34	05	08	5	171	106	084	36	16	12	12	27	10	26	06	110
106	4	3	16	1	12	08	07	2	194	104	093	46	13	15	13	47	03	36	24	109
107	4	3	17	1	12	07	04	1	214	109	098	43	14	18	11	36	12	36	04	098
108	4	3	17	1	35	07	01	4	131	080	070	41	09	13	12	29	04	11	04	092
109	4	3	16	1	36	07	09	4	145	071	085	30	16	12	09	15	05	25	17	098
110	4	3	17	1	01	07	01	5	162	085	077	39	10	09	09	23	03	14	10	106
111	3	3	17	1	01	06	07	2	169	095	001	44	15	18	12	00	09	22	39	103
112	2	3	18	1	05	09	06	5	109	067	067	15	13	18	09	19	06	16	19	111
113	4	3	17	1	18	07	06	2	123	061	062	14	05	09	05	13	11	24	13	102
114	4	3	17	1	04	06	08	2	165	083	073	34	14	10	05	20	05	21	06	102
115	4	3	18	1	05	06	08	5	120	065	079	20	08	10	09	24	09	19	14	099
116	2	3	17	1	29	06	01	5	106	045	076	25	08	13	07	25	04	18	20	088
117	3	3	17	1	05	06	04	1	213	093	091	26	09	18	12	39	08	33	11	105
118	2	3	17	1	11	06	01	5	135	064	078	22	12	12	09	19	02	04	01	094
119	3	3	17	1	01	05	05	1	122	063	087	39	15	11	10	30	06	13	06	108
120	3	3	18	1	34	05	01	4	170	087	090	36	13	13	11	22	08	11	06	092
121	2	4	17	1	31	07	01	5	089	041	067	19	07	15	07	12	03	13	10	084
122	2	3	18	1	05	07	04	2	185	093	093	45	15	15	10	44	07	28	13	098
123	3	3	19	1	08	07	05	1	193	094	084	36	10	16	12	38	10	27	13	110
124	3	3	18	1	36	07	01	5	182	082	077	38	09	12	06	19	10	22	13	103
125	3	3	17	1	12	08	07	1	223	112	106	44	18	20	14	48	05	34	19	112
126	4	3	17	1	26	06	01	5	151	069	086	36	10	16	11	16	03	22	14	095
127	4	3	18	1	20	07	09	3	191	106	091	42	11	19	11	31	09	29	14	088
128	2	3	17	1	36	06	01	4	137	070	089	27	08	13	10	26	12	13	17	096
129	2	3	17	1	05	07	01	4	151	075	094	34	12	13	10	23	04	19	13	096
130	1	3	17	1	17	08	02	2	168	079	093	40	16	16	12	39	05	25	02	098
131	4	3	17	1	36	07	01	5	124	062	069	26	09	15	09	25	03	10	11	089
132	2	4	17	1	05	07	01	5	160	070	076	23	07	18	11	26	10	21	14	089
133	2	4	17	1	11	09	08	1	151	086	085	34	04	13	10	31	10	11	13	104
134	2	4	17	1	19	08	01	4	136	065	077	27	08	14	09	20	09	04	00	088
135	2	4	17	1	19	06	08	3	159	091	089	40	08	07	10	24	11	18	26	099
136	2	4	18	1	05	06	07	1	183	097	092	46	15	19	12	36	10	33	19	101
137	3	4	18	1	10	05	07	2	130	080	079	31	07	13	13	20	08	19	04	101
138	4	4	18	1	14	06	07	1	217	110	101	47	15	16	13	39	04	27	14	113
139	4	4	17	1	13	07	05	1	204	108	095	45	12	13	12	28	07	26	24	106
140	2	4	17	1	01	06	01	5	116	040	063	19	04	07	08	15	05	19	14	091
141	2	4	19	1	05	05	04	4	132	071	080	28	09	10	09	13	07	15	11	092
142	4	4	17	1	12	07	04	2	233	115	096	45	18	20	10	52	05	38	01	115
143	3	4	17	1	01	07	05	1	161	085	086	33	06	13	11	29	06	16	25	111
144	2	4	18	1	14	09	07	3	184	102	084	45	14	20	15	27	04	20	21	088

ID	1	2	3	4	5	6	7	8	9	10	11	12	13	14	15	16	17	18	19	20
145	2	4	17	1	30	07	01	5	132	078	063	22	08	10	09	20	08	04	01	085
146	2	4	17	1	29	07	01	5	112	063	076	23	03	08	07	12	12	12	27	091
147	3	5	17	1	19	08	08	3	152	088	090	35	09	09	07	20	09	19	19	104
148	3	5	18	1	10	06	08	3	149	074	081	43	12	14	10	31	09	26	14	103
149	2	5	18	1	31	06	03	2	165	079	087	36	16	20	08	29	09	26	04	102
150	3	5	18	1	19	08	05	2	121	079	086	24	06	09	11	12	10	19	13	107
151	3	5	17	1	08	07	05	2	164	068	083	36	11	14	12	24	05	18	04	107
152	2	5	17	1	05	06	04	2	162	091	090	35	15	20	10	36	10	39	24	100
153	2	5	18	1	04	08	08	3	117	058	065	17	05	09	02	16	06	16	23	100
154	3	5	17	1	19	06	09	3	100	052	071	17	09	13	09	15	10	26	23	092
155	3	5	18	1	36	06	01	5	103	042	064	12	04	06	04	11	03	16	27	085
156	1	5	17	1	05	12	04	4	186	093	077	30	08	18	06	20	02	31	20	086
157	2	5	17	1	19	08	07	1	134	078	085	35	07	11	10	22	11	08	00	092
158	2	5	19	1	05	06	07	2	124	066	085	29	06	11	02	18	08	33	20	078
159	3	5	18	1	36	11	01	4	092	047	065	16	06	04	04	13	11	13	10	095
160	3	5	18	1	36	09	01	5	123	070	081	23	07	13	09	21	04	14	16	108
161	2	5	17	1	10	06	07	2	172	089	092	39	16	15	11	34	09	36	17	103
162	3	5	17	1	01	03	08	1	167	103	100	46	06	09	11	37	07	21	23	092
163	3	5	17	1	05	08	04	1	164	087	088	39	11	16	10	38	11	26	11	103
164	3	5	18	1	34	07	01	3	113	061	072	16	09	17	07	10	06	26	21	084
165	3	5	17	1	05	12	05	1	149	083	095	36	09	11	04	24	12	33	24	099
166	4	5	18	1	21	09	09	1	140	027	063	14	07	05	05	20	07	18	10	086
167	2	5	17	1	05	08	01	5	111	062	086	28	15	11	08	19	09	19	03	096
168	4	5	17	1	18	07	08	3	172	102	101	43	08	17	13	39	11	11	04	109
169	4	5	18	1	31	06	01	5	131	073	079	31	07	10	12	15	03	09	01	083
170	3	5	18	1	05	06	01	5	130	074	084	19	07	12	05	17	06	22	13	087
171	2	5	18	1	04	07	07	1	156	086	091	28	11	12	10	28	11	23	17	106
172	3	5	17	1	02	07	02	3	110	062	075	31	07	14	06	20	07	19	10	100
173	2	5	19	1	36	07	08	4	162	074	079	30	07	14	12	17	10	16	01	094
174	3	5	18	1	22	05	09	3	121	075	081	30	05	11	09	17	02	00	01	085
175	2	5	17	1	12	05	04	3	177	094	082	40	16	15	08	25	04	37	24	092
176	3	5	17	1	19	06	09	5	092	048	070	12	11	08	02	11	10	25	10	094
177	3	5	17	1	13	08	07	1	198	102	092	46	14	14	10	38	09	31	06	100
178	2	5	19	1	17	08	01	4	088	041	066	14	05	09	02	09	03	09	00	077
179	3	5	17	1	36	07	07	4	134	077	081	39	05	18	12	20	05	13	09	082
180	4	5	16	1	18	05	01	5	147	084	085	25	09	12	09	15	08	06	03	095
181	4	5	17	1	23	05	07	1	164	086	083	31	09	10	11	31	10	38	04	108
182	4	5	17	1	10	06	08	3	205	105	094	34	15	16	15	37	07	33	11	095
183	2	5	17	1	19	07	07	1	170	087	098	35	13	18	09	25	06	05	17	100
184	2	5	17	1	36	07	01	5	156	090	077	40	09	08	08	20	08	08	03	080
185	2	5	18	1	05	08	07	1	221	113	103	47	17	19	15	46	08	36	19	111
186	2	5	18	1	25	06	01	4	131	053	086	19	13	12	10	18	10	30	31	083
187	3	5	16	1	36	07	07	3	142	087	079	35	09	11	09	18	10	29	03	114
188	2	5	18	1	11	08	05	1	163	097	096	46	06	13	12	24	10	04	06	094
189	2	5	18	1	22	08	03	1	128	073	052	40	09	14	07	12	10	21	17	092
190	2	5	17	1	19	07	08	1	135	070	083	31	10	14	10	22	10	18	27	100
191	3	5	18	1	36	07	01	5	116	057	074	22	07	14	10	22	06	13	00	096
192	3	5	18	1	34	06	01	4	147	071	083	31	14	13	09	20	02	18	17	093

ID	1	2	3	4	5	6	7	8	9	10	11	12	13	14	15	16	17	18	19	20
193	2	6	18	1	25	07	04	2	201	098	092	42	13	19	13	42	04	27	16	104
194	2	6	16	1	07	09	07	1	148	073	082	31	10	15	07	25	12	32	16	105
195	3	6	17	1	05	09	04	1	203	100	095	44	14	17	08	43	10	30	09	096
196	4	6	18	1	13	06	06	1	161	064	068	04	05	19	12	30	03	26	07	104
197	4	6	17	1	15	07	09	3	144	085	081	41	13	13	11	16	04	06	07	103
198	3	6	17	1	29	05	08	3	107	055	082	25	08	14	11	19	07	13	21	081
199	3	6	18	1	36	08	09	1	163	086	087	32	12	13	10	29	04	30	18	110
200	2	6	18	1	04	09	04	2	156	087	096	42	16	10	08	31	10	29	06	114
201	1	6	19	1	03	06	05	5	078	035	043	18	01	04	01	08	04	22	30	097
202	3	6	18	1	18	05	07	1	200	100	097	45	16	18	12	36	05	33	13	108
203	1	6	18	1	05	07	04	3	187	089	073	36	10	19	10	30	07	31	16	105
204	2	6	17	1	31	08	01	5	141	084	081	38	10	12	07	20	08	11	07	107
205	4	6	18	1	19	07	04	1	187	088	096	43	10	19	12	37	09	29	16	105
206	4	6	17	1	04	06	07	1	150	077	074	40	10	08	07	20	07	31	14	106
207	2	6	18	1	26	07	01	5	104	067	075	28	03	08	07	15	04	13	18	076
208	3	5	18	1	34	08	08	4	120	059	081	30	02	08	07	23	04	13	19	081
209	2	7	17	1	25	08	01	4	191	088	091	39	15	17	11	26	02	11	04	088
210	1	7	17	1	22	06	01	5	113	053	069	25	08	07	04	21	01	17	06	095
211	4	7	17	1	10	04	07	3	154	081	091	36	13	14	09	24	02	15	10	108
212	4	7	17	1	18	06	06	4	119	071	087	35	08	08	08	13	06	18	20	083
213	4	7	17	1	34	08	01	5	131	048	061	20	09	10	06	16	00	14	06	094
214	3	7	17	1	05	07	04	1	140	062	072	24	09	14	10	27	02	23	14	107
215	3	7	17	1	05	06	07	2	176	081	090	43	18	15	10	36	07	20	14	098
216	3	7	17	1	01	06	08	4	173	105	095	43	16	16	12	40	10	22	36	100
217	3	8	18	1	12	05	07	2	214	103	097	45	11	18	14	48	09	38	06	111
218	3	8	17	1	18	06	06	3	176	078	083	36	11	15	14	12	08	10	00	096
219	3	8	18	1	07	08	09	2	139	089	090	35	09	14	13	21	09	04	01	109
220	2	8	18	1	07	06	02	2	098	038	091	28	07	05	09	15	09	08	07	086
221	3	8	17	1	05	07	08	1	175	089	080	34	10	11	08	25	03	11	06	099
222	2	8	18	1	36	09	01	5	190	102	073	44	09	14	07	20	01	03	00	104
223	3	8	17	1	06	07	02	1	178	084	089	40	11	16	13	39	08	36	23	096
224	1	8	18	1	23	06	09	1	143	072	085	33	08	14	11	23	05	20	00	097
225	4	8	18	1	20	07	05	2	134	071	080	34	13	17	10	18	07	16	25	094
226	4	8	17	1	36	09	07	2	194	102	081	41	15	16	11	22	11	20	10	107
227	4	8	18	1	18	06	01	4	109	067	080	29	08	07	11	10	07	05	06	082
228	4	8	18	1	35	06	01	5	105	064	057	21	07	11	09	15	02	19	11	088
229	4	8	17	1	36	04	01	5	139	073	079	16	04	14	09	15	06	10	07	098
230	4	8	17	1	01	06	09	2	125	078	083	41	12	12	11	24	10	17	07	101
231	2	8	17	1	26	07	09	3	178	074	080	23	08	19	12	29	05	24	14	098
232	4	8	17	1	08	05	07	1	053	032	040	07	03	06	02	12	02	16	07	099
233	4	7	18	1	02	06	04	1	216	113	099	46	16	17	11	45	09	36	21	114
234	2	2	17	1	34	08	01	5	184	079	094	40	14	17	14	28	06	19	10	079

ID	1	2	3	4	5	6	7	8	9	10	11	12	13	14	15	16	17	18	19	20
145	2	4	17	1	30	07	01	5	132	078	063	22	08	10	09	20	08	04	01	085
146	2	4	17	1	29	07	01	5	112	063	076	23	03	08	07	12	12	12	27	091
147	3	5	17	1	19	08	08	3	152	088	090	35	09	09	07	20	09	19	19	104
148	3	5	18	1	10	06	08	3	149	074	081	43	12	14	10	31	09	26	14	103
149	2	5	18	1	31	06	03	2	165	079	087	36	16	20	08	29	09	26	04	102
150	3	5	18	1	19	08	05	2	121	079	086	24	06	09	11	12	10	19	13	107
151	3	5	17	1	08	07	05	2	164	068	083	36	11	14	12	24	05	18	04	107
152	2	5	17	1	05	06	04	2	162	091	090	35	15	20	10	36	10	39	24	100
153	2	5	18	1	04	08	08	3	117	058	065	17	05	09	02	16	06	16	23	100
154	3	5	17	1	19	06	09	3	100	052	071	17	09	13	09	15	10	26	23	092
155	3	5	18	1	36	06	01	5	103	042	064	12	04	06	04	11	03	16	27	085
156	1	5	17	1	05	12	04	4	186	093	077	30	08	18	06	20	02	31	20	086
157	2	5	17	1	19	08	07	1	134	078	085	35	07	11	10	22	11	08	00	092
158	2	5	19	1	05	06	07	2	124	066	085	29	06	11	02	18	08	33	20	078
159	3	5	18	1	36	11	01	4	092	047	065	16	06	04	04	13	11	13	10	095
160	3	5	18	1	36	09	01	5	123	070	081	23	07	13	09	21	04	14	16	108
161	2	5	17	1	10	06	07	2	172	089	092	39	16	15	11	34	09	36	17	103
162	3	5	17	1	01	03	08	1	167	103	100	46	06	09	11	37	07	21	23	092
163	3	5	17	1	05	08	04	1	164	087	088	39	11	16	10	38	11	26	11	103
164	3	5	18	1	34	07	01	3	113	061	072	16	09	17	07	10	06	26	21	084
165	3	5	17	1	05	12	05	1	149	083	095	36	09	11	04	24	12	33	24	099
166	4	5	18	1	21	09	09	1	140	027	063	14	07	05	05	20	07	18	10	086
167	2	5	17	1	05	08	01	5	111	062	086	28	15	11	08	19	09	19	03	096
168	4	5	17	1	18	07	08	3	172	102	101	43	08	17	13	39	11	11	04	109
169	4	5	18	1	31	06	01	5	131	073	079	31	07	10	12	15	03	09	01	083
170	3	5	18	1	05	06	01	5	130	074	084	19	07	12	05	17	06	22	13	087
171	2	5	18	1	04	07	07	1	156	086	091	28	11	12	10	28	11	23	17	106
172	3	5	17	1	02	07	02	3	110	062	075	31	07	14	06	20	07	19	10	100
173	2	5	19	1	36	07	08	4	162	074	079	30	07	14	12	17	10	16	01	094
174	3	5	18	1	22	05	09	3	121	075	081	30	05	11	09	17	02	00	01	085
175	2	5	17	1	12	05	04	3	177	094	082	40	16	15	08	25	04	37	24	092
176	3	5	17	1	19	06	09	5	092	048	070	12	11	08	02	11	10	25	10	094
177	3	5	17	1	13	08	07	1	198	102	092	46	14	14	10	38	09	31	06	100
178	2	5	19	1	17	08	01	4	088	041	066	14	05	09	02	09	03	09	00	077
179	3	5	17	1	36	07	07	4	134	077	081	39	05	18	12	20	05	13	09	082
180	4	5	16	1	18	05	01	5	147	084	085	25	09	12	09	15	08	06	03	095
181	4	5	17	1	23	05	07	1	164	086	083	31	09	10	11	31	10	38	04	108
182	4	5	17	1	10	06	08	3	205	105	094	34	15	16	15	37	07	33	11	095
183	2	5	17	1	19	07	07	1	170	087	098	35	13	18	09	25	06	05	17	100
184	2	5	17	1	36	07	01	5	156	090	077	40	09	08	08	20	08	08	03	080
185	2	5	18	1	05	08	07	1	221	113	103	47	17	19	15	46	08	36	19	111
186	2	5	18	1	25	06	01	4	131	053	086	19	13	12	10	18	10	30	31	083
187	3	5	16	1	36	07	07	3	142	087	079	35	09	11	09	18	10	29	03	114
188	2	5	18	1	11	08	05	1	163	097	096	46	06	13	12	24	10	04	06	094
189	2	5	18	1	22	08	03	1	128	073	052	40	09	14	07	12	10	21	17	092
190	2	5	17	1	19	07	08	1	135	070	083	31	10	14	10	22	10	18	27	100
191	3	5	18	1	36	07	01	5	116	057	074	22	07	14	10	22	06	13	00	096
192	3	5	18	1	34	06	01	4	147	071	083	31	14	13	09	20	02	18	17	093

ID	1	2	3	4	5	6	7	8	9	10	11	12	13	14	15	16	17	18	19	20
193	2	6	18	1	25	07	04	2	201	098	092	42	13	19	13	42	04	27	16	104
194	2	6	16	1	07	09	07	1	148	073	082	31	10	15	07	25	12	32	16	105
195	3	6	17	1	05	09	04	1	203	100	095	44	14	17	08	43	10	30	09	096
196	4	6	18	1	13	06	06	1	161	064	068	04	05	19	12	30	03	26	07	104
197	4	6	17	1	15	07	09	3	144	085	081	41	13	13	11	16	04	06	07	103
198	3	6	17	1	29	05	08	3	107	055	082	25	08	14	11	19	07	13	21	081
199	3	6	18	1	36	08	09	1	163	086	087	32	12	13	10	29	04	30	18	110
200	2	6	18	1	04	09	04	2	156	087	096	42	16	10	08	31	10	29	06	114
201	1	6	19	1	03	06	05	5	078	035	043	18	01	04	01	08	04	22	30	097
202	3	6	18	1	18	05	07	1	200	100	097	45	16	18	12	36	05	33	13	108
203	1	6	18	1	05	07	04	3	187	089	073	36	10	19	10	30	07	31	16	105
204	2	6	17	1	31	08	01	5	141	084	081	38	10	12	07	20	08	11	07	107
205	4	6	18	1	19	07	04	1	187	088	096	43	10	19	12	37	09	29	16	105
206	4	6	17	1	04	06	07	1	150	077	074	40	10	08	07	20	07	31	14	106
207	2	6	18	1	26	07	01	5	104	067	075	28	03	08	07	15	04	13	18	076
208	3	5	18	1	34	08	08	4	120	059	081	30	02	08	07	23	04	13	19	081
209	2	7	17	1	25	08	01	4	191	088	091	39	15	17	11	26	02	11	04	088
210	1	7	17	1	22	06	01	5	113	053	069	25	08	07	04	21	01	17	06	095
211	4	7	17	1	10	04	07	3	154	081	091	36	13	14	09	24	02	15	10	108
212	4	7	17	1	18	06	06	4	119	071	087	35	08	08	08	13	06	18	20	083
213	4	7	17	1	34	08	01	5	131	048	061	20	09	10	06	16	00	14	06	094
214	3	7	17	1	05	07	04	1	140	062	072	24	09	14	10	27	02	23	14	107
215	3	7	17	1	05	06	07	2	176	081	090	43	18	15	10	36	07	20	14	098
216	3	7	17	1	01	06	08	4	173	105	095	43	16	16	12	40	10	22	36	100
217	3	8	18	1	12	05	07	2	214	103	097	45	11	18	14	48	09	38	06	111
218	3	8	17	1	18	06	06	3	176	078	083	36	11	15	14	12	08	10	00	096
219	3	8	18	1	07	08	09	2	139	089	090	35	09	14	13	21	09	04	01	109
220	2	8	18	1	07	06	02	2	098	038	091	28	07	05	09	15	09	08	07	086
221	3	8	17	1	05	07	08	1	175	089	080	34	10	11	08	25	03	11	06	099
222	2	8	18	1	36	09	01	5	190	102	073	44	09	14	07	20	01	03	00	104
223	3	8	17	1	06	07	02	1	178	084	089	40	11	16	13	39	08	36	23	096
224	1	8	18	1	23	06	09	1	143	072	085	33	08	14	11	23	05	20	00	097
225	4	8	18	1	20	07	05	2	134	071	080	34	13	17	10	18	07	16	25	094
226	4	8	17	1	36	09	07	2	194	102	081	41	15	16	11	22	11	20	10	107
227	4	8	18	1	18	06	01	4	109	067	080	29	08	07	11	10	07	05	06	082
228	4	8	18	1	35	06	01	5	105	064	057	21	07	11	09	15	02	19	11	088
229	4	8	17	1	36	04	01	5	139	073	079	16	04	14	09	15	06	10	07	098
230	4	8	17	1	01	06	09	2	125	078	083	41	12	12	11	24	10	17	07	101
231	2	8	17	1	26	07	09	3	178	074	080	23	08	19	12	29	05	24	14	098
232	4	8	17	1	08	05	07	1	053	032	040	07	03	06	02	12	02	16	07	099
233	4	7	18	1	02	06	04	1	216	113	099	46	16	17	11	45	09	36	21	114
234	2	2	17	1	34	08	01	5	184	079	094	40	14	17	14	28	06	19	10	079

Table B.3 Project TALENT Data—20 Variables for 271 Twelfth-Grade Females

ID	1	2	3	4	5	6	7	8	9	10	11	12	13	14	15	16	17	18	19	20
1	2	5	16	2	13	04	07	3	142	083	094	39	06	04	09	15	06	31	15	092
2	2	3	18	2	32	05	01	5	148	095	094	39	09	09	06	24	10	05	11	089
3	3	1	18	2	24	07	08	4	136	073	092	37	14	18	11	29	04	14	20	097
4	2	1	17	2	36	04	05	1	207	100	108	46	10	11	14	43	06	18	17	118
5	3	1	17	2	29	03	01	5	100	047	079	15	13	05	05	27	08	05	39	096
6	3	1	17	2	24	05	08	5	124	059	093	46	09	05	11	28	11	16	36	102
7	4	1	19	2	29	12	01	4	091	047	066	11	06	07	05	12	09	06	17	100
8	4	1	18	2	06	03	02	1	140	093	089	42	08	09	12	32	05	11	23	096
9	4	1	19	2	32	05	01	5	148	073	000	42	08	14	12	00	06	24	24	089
10	4	1	18	2	27	03	01	5	056	034	086	23	04	04	08	15	04	13	37	095
11	4	1	17	2	06	05	07	1	121	066	095	40	10	09	11	35	07	06	29	107
12	2	1	17	2	27	03	08	2	084	071	084	37	06	06	09	11	06	03	15	100
13	2	1	17	2	27	04	08	1	152	090	101	43	13	15	12	40	11	14	11	111
14	2	1	18	2	27	04	01	5	086	060	087	28	06	06	06	07	09	07	36	087
15	3	1	17	2	32	03	01	5	101	059	093	33	14	09	09	27	10	01	36	085
16	3	1	16	2	36	04	01	5	110	062	078	23	09	10	12	13	06	17	27	108
17	3	1	17	2	15	03	07	3	117	082	098	45	06	09	09	20	07	05	26	090
18	3	1	17	2	07	05	05	2	180	083	096	42	13	13	13	45	10	26	24	101
19	4	2	16	2	27	05	01	5	146	087	095	41	08	13	12	22	12	29	34	092
20	4	2	17	2	27	05	01	5	118	080	081	43	09	07	09	15	07	06	13	094
21	4	2	17	2	06	05	05	1	167	094	088	40	13	09	11	28	07	27	26	106
22	4	2	17	2	36	05	05	1	122	075	084	24	11	07	09	11	08	21	31	103
23	4	2	17	2	27	06	01	5	086	052	088	17	03	02	05	12	07	02	30	091
24	4	2	16	2	28	04	05	1	199	094	102	45	13	11	11	37	10	14	23	107
25	4	2	17	2	27	10	01	4	116	086	084	25	12	10	07	18	02	17	30	091
26	4	2	17	2	36	03	08	1	103	058	080	27	05	06	06	13	10	11	19	088
27	4	2	17	2	24	05	01	4	097	042	085	24	05	08	11	19	10	16	17	093
28	4	2	17	2	27	04	07	5	111	071	081	32	05	07	09	16	11	33	37	110
29	4	2	17	2	36	06	07	1	152	092	091	39	11	09	07	34	07	25	11	119
30	4	2	16	2	15	05	09	4	133	077	093	32	09	04	15	24	11	01	17	104
31	4	2	17	2	27	03	01	5	073	055	083	32	03	04	03	13	09	04	37	099
32	2	2	17	2	07	05	02	1	173	094	104	43	13	17	14	24	10	29	31	077
33	3	2	17	2	36	04	07	1	207	103	098	43	15	18	13	46	12	29	29	116
34	2	2	18	2	27	05	01	5	144	092	099	39	10	12	10	25	08	14	37	101
35	3	2	18	2	25	04	01	5	117	059	086	26	02	04	14	15	12	11	25	102
36	3	2	17	2	24	04	01	5	163	094	091	43	12	08	09	29	03	11	03	095
37	3	2	17	2	06	06	02	1	189	098	107	42	17	13	12	39	06	22	01	097
38	2	2	17	2	15	07	05	1	182	097	094	41	15	10	14	41	09	16	30	108
39	3	2	18	2	27	06	01	5	098	063	087	28	06	10	10	20	05	03	27	099
40	3	5	17	2	32	05	01	5	172	098	103	45	15	14	11	31	05	04	24	110
41	3	5	17	2	27	04	08	5	110	054	092	30	09	12	09	19	07	13	30	099
42	4	2	17	2	23	04	01	5	182	093	099	42	10	15	13	39	10	31	13	094
43	4	2	17	2	06	06	02	2	185	099	104	42	11	13	09	33	08	17	21	101
44	4	2	17	2	24	04	01	5	125	059	079	28	08	10	12	27	08	13	19	092
45	3	2	16	2	27	06	01	5	105	060	072	14	12	04	09	07	10	06	39	096
46	3	2	17	2	27	07	01	5	036	021	063	15	04	08	04	09	07	03	27	100
47	4	2	17	2	27	05	01	4	153	085	094	32	11	11	08	22	08	12	29	093
48	4	2	17	2	06	04	02	1	145	081	084	31	04	05	09	27	10	14	11	108

ID	1	2	3	4	5	6	7	8	9	10	11	12	13	14	15	16	17	18	19	20
49	4	2	17	2	07	05	05	1	171	097	096	44	13	14	12	38	07	24	23	097
50	4	2	17	2	01	07	01	5	123	063	082	31	12	07	10	23	04	08	40	094
51	4	2	17	2	27	04	01	5	097	055	080	21	04	11	07	18	07	18	30	094
52	3	2	17	2	36	04	01	5	164	101	096	45	13	15	14	37	10	18	19	096
53	3	2	17	2	27	04	01	5	099	068	082	32	08	08	08	05	11	01	33	105
54	2	2	17	2	27	04	01	5	109	064	074	27	09	07	01	13	10	04	34	087
55	2	2	18	2	36	07	08	1	141	062	096	37	06	16	11	39	05	24	16	104
56	3	2	17	2	27	06	01	4	178	095	102	44	09	11	12	27	09	18	33	090
57	3	2	17	2	27	04	08	1	129	081	096	44	11	09	12	30	12	07	31	106
58	3	2	17	2	07	04	07	1	147	091	089	32	07	11	08	30	10	21	24	098
59	3	2	17	2	27	04	01	5	138	084	093	44	10	14	09	25	11	04	16	111
60	3	2	18	2	29	05	01	5	124	080	094	38	14	10	09	21	09	04	23	086
61	3	2	18	2	27	05	01	4	120	077	084	42	09	07	08	16	04	18	31	101
62	3	2	16	2	27	06	01	5	100	083	093	33	08	10	09	17	08	15	27	087
63	4	2	17	2	24	06	09	5	109	051	081	28	04	04	08	18	08	24	29	089
64	4	2	16	2	24	09	07	2	154	079	095	42	13	12	09	35	08	29	26	098
65	4	2	18	2	24	05	08	1	081	045	066	17	04	06	04	14	10	24	21	091
66	3	2	17	2	06	05	02	1	182	097	103	37	08	09	12	24	11	11	10	111
67	3	3	18	2	06	06	07	1	174	101	102	42	09	11	10	34	06	09	20	116
68	3	3	17	2	27	10	01	5	121	078	086	30	09	08	06	21	07	00	26	097
69	3	3	17	2	27	05	07	1	153	086	091	42	11	07	08	17	06	04	24	102
70	2	3	17	2	27	05	01	4	130	077	103	33	05	09	10	19	08	13	39	102
71	2	3	17	2	29	06	08	5	093	057	079	24	11	08	07	16	07	14	30	085
72	3	3	18	2	06	05	07	1	205	103	111	46	15	14	13	47	03	18	24	102
73	4	3	17	2	23	05	08	5	143	071	097	37	11	11	09	31	09	01	36	103
74	4	3	18	2	27	05	01	5	167	093	088	35	12	13	07	11	10	13	19	102
75	2	3	17	2	24	03	01	5	128	068	087	36	11	12	15	26	07	19	23	099
76	2	3	19	2	27	05	01	4	112	077	077	35	08	06	10	18	07	07	15	104
77	2	3	17	2	29	05	01	5	132	077	094	40	02	11	08	21	05	12	36	097
78	4	3	18	2	36	06	08	1	113	078	083	37	12	07	11	17	10	21	29	085
79	4	3	16	2	13	05	07	5	073	043	071	16	03	02	09	07	09	08	00	108
80	3	3	17	2	07	06	07	1	178	101	099	41	10	14	13	25	06	12	09	099
81	3	3	18	2	17	07	08	4	174	106	098	43	16	13	11	29	07	09	17	108
82	3	3	18	2	36	03	01	5	103	062	081	24	07	12	13	19	05	05	33	108
83	3	3	17	2	27	05	01	4	160	087	109	45	17	14	13	33	02	21	33	098
84	3	3	18	2	07	05	05	2	137	078	107	42	10	07	11	42	06	13	27	094
85	3	3	17	2	27	05	01	5	138	071	095	31	07	04	10	16	10	03	24	095
86	3	3	17	2	24	05	07	1	184	086	093	37	12	09	11	44	08	18	10	116
87	2	3	17	2	03	04	05	1	201	094	110	47	19	17	14	51	03	34	10	104
88	3	3	18	2	27	06	01	5	120	060	084	27	08	08	10	26	07	09	29	097
89	3	3	18	2	01	05	01	5	111	067	080	26	08	06	04	15	03	00	21	091
90	3	3	17	2	33	05	01	5	148	078	108	39	07	06	10	25	08	12	29	090
91	3	3	17	2	27	05	02	3	099	060	079	38	08	06	12	19	02	08	33	094
92	3	3	17	2	27	05	01	4	175	086	100	40	13	08	07	23	04	16	24	102
93	4	3	17	2	33	03	06	4	084	045	081	27	11	08	14	17	11	07	29	097
94	4	3	17	2	18	07	05	4	109	066	079	28	09	09	10	25	04	06	34	090
95	3	3	17	2	07	05	02	1	147	078	096	40	09	09	08	17	05	03	33	114
96	4	3	17	2	24	04	07	2	134	068	088	34	09	08	09	21	11	13	21	115

ID	1	2	3	4	5	6	7	8	9	10	11	12	13	14	15	16	17	18	19	20
97	4	3	17	2	15	06	01	5	142	088	075	37	04	11	00	11	05	12	24	099
98	2	3	17	2	27	05	07	1	153	090	087	41	08	12	07	16	09	23	39	103
99	3	3	17	2	27	04	01	5	102	061	085	30	07	07	08	20	09	04	23	089
100	3	3	17	2	27	04	01	4	109	057	082	22	06	04	06	13	11	05	36	105
101	4	3	18	2	27	04	01	5	110	069	078	42	10	12	09	20	08	13	37	097
102	4	3	18	2	32	06	01	5	128	080	077	28	15	10	13	25	08	07	30	101
103	4	3	17	2	06	05	07	2	207	108	102	48	19	11	13	41	06	14	09	092
104	3	3	17	2	18	05	01	5	105	069	087	24	04	05	04	19	08	01	06	096
105	3	3	18	2	32	05	01	4	153	080	100	38	15	10	10	22	07	03	36	104
106	2	3	17	2	29	04	09	4	092	052	082	18	04	04	10	14	10	06	19	087
107	2	3	17	2	32	05	01	5	108	070	082	24	07	11	12	29	10	12	37	100
108	4	3	17	2	32	05	07	1	134	077	080	36	11	04	05	19	11	07	06	117
109	4	3	17	2	06	05	07	1	139	092	096	45	13	06	09	22	11	11	23	120
110	4	3	17	2	06	04	07	1	117	079	076	40	08	08	08	20	03	08	26	101
111	4	3	18	2	36	05	05	5	089	055	081	16	05	08	08	11	08	14	24	093
112	3	3	18	2	32	04	01	5	123	077	084	36	06	05	11	10	04	15	29	101
113	2	3	17	2	28	05	01	5	129	074	093	44	09	12	10	28	03	11	17	091
114	3	3	18	2	36	05	01	5	108	054	074	20	09	11	06	16	08	11	26	093
115	4	3	17	2	07	04	07	2	172	081	099	46	08	15	13	41	08	25	03	095
116	4	3	17	2	07	05	02	1	199	090	088	37	13	11	10	39	05	18	20	107
117	4	3	18	2	15	05	07	4	105	063	075	29	05	02	04	10	09	08	24	097
118	4	3	18	2	24	06	07	1	132	074	086	35	07	06	07	22	06	15	07	087
119	4	3	17	2	19	04	01	4	103	053	081	24	07	10	11	14	09	13	34	088
120	2	3	17	2	27	04	08	5	160	083	100	46	10	12	11	23	07	13	23	098
121	2	2	17	2	07	05	02	1	171	090	099	47	18	14	14	23	06	06	19	109
122	3	1	17	2	24	07	08	1	153	078	096	00	00	00	00	36	09	17	17	098
123	3	1	17	2	27	06	01	4	115	058	084	30	11	13	08	27	11	13	36	093
124	2	3	17	2	07	05	05	1	163	086	104	41	15	09	11	43	10	13	33	089
125	3	3	17	2	23	06	08	4	141	087	087	40	14	10	10	20	07	11	30	105
126	3	3	18	2	27	04	08	4	145	076	081	31	10	11	12	19	09	21	31	105
127	3	3	17	2	32	05	01	5	111	073	086	26	08	08	08	18	08	05	39	097
128	4	3	17	2	17	03	01	5	124	079	098	39	05	12	10	23	09	16	37	095
129	2	3	17	2	27	07	01	5	147	073	099	38	09	08	13	24	07	09	40	100
130	2	3	17	2	27	05	01	5	120	074	091	36	09	10	09	16	04	04	30	091
131	3	3	17	2	19	06	09	3	104	061	092	37	07	14	09	23	03	08	39	109
132	2	4	17	2	27	03	01	5	127	072	087	31	04	02	09	23	06	14	30	091
133	2	4	17	2	07	04	07	1	162	089	099	44	15	10	13	25	07	04	26	096
134	3	4	17	2	06	06	09	1	152	067	098	33	10	13	10	26	05	07	21	102
135	3	4	17	2	36	03	01	5	093	058	077	25	10	07	06	13	03	08	33	092
136	3	4	18	2	27	05	01	5	084	061	076	26	05	07	10	20	06	04	33	097
137	3	4	17	2	27	05	05	1	116	068	088	21	04	06	07	11	11	04	36	102
138	3	4	17	2	06	04	02	1	138	065	086	36	08	08	10	20	04	04	14	109
139	2	4	18	2	15	06	05	1	151	000	086	35	16	06	03	28	01	01	03	105
140	2	4	17	2	27	06	01	4	125	069	080	24	06	07	10	19	10	04	30	084
141	2	4	18	2	27	05	01	5	108	063	090	21	10	06	09	23	02	03	34	084
142	2	4	17	2	06	04	05	1	137	069	090	33	09	10	09	20	11	13	27	110
143	2	4	17	2	32	04	01	5	102	059	081	31	07	05	10	16	09	07	26	095
144	2	4	17	2	32	04	01	4	140	074	090	32	08	07	10	28	09	09	26	080

ID	1	2	3	4	5	6	7	8	9	10	11	12	13	14	15	16	17	18	19	20
145	4	4	17	2	27	05	05	4	138	081	090	38	08	08	09	25	04	16	34	106
146	3	4	17	2	06	06	02	1	166	094	090	39	14	12	10	25	10	08	16	104
147	2	4	18	2	27	04	01	4	127	080	089	37	14	12	13	16	10	15	27	107
148	3	4	18	2	16	04	05	2	184	106	099	44	11	15	11	37	07	24	26	106
149	4	4	17	2	32	04	08	3	131	078	075	27	04	08	09	20	03	07	30	106
150	4	4	17	2	18	06	06	2	158	097	084	44	09	07	09	31	04	16	16	113
151	3	4	17	2	32	06	01	5	131	083	092	36	12	10	08	31	00	12	36	104
152	2	4	18	2	32	03	01	5	086	047	080	21	07	07	10	16	10	00	29	088
153	2	4	17	2	29	06	01	4	100	042	066	36	10	07	13	15	08	03	17	074
154	2	4	17	2	17	05	01	4	112	058	090	36	07	08	10	11	00	01	16	083
155	2	4	18	2	01	06	02	2	178	088	105	45	15	16	15	39	01	08	37	098
156	4	4	17	2	29	05	01	5	098	060	075	29	08	05	11	10	06	10	19	096
157	4	4	17	2	27	06	01	5	108	068	099	29	08	10	13	23	06	05	33	084
158	3	4	17	2	27	03	07	2	155	089	100	42	10	09	14	21	10	17	39	096
159	3	4	17	2	25	05	05	5	109	071	081	20	03	06	08	17	11	00	11	113
160	3	5	17	2	36	06	08	2	187	110	067	48	13	12	13	37	07	10	00	112
161	3	5	18	2	27	04	01	5	136	070	094	42	10	12	12	20	07	07	33	096
162	2	5	18	2	27	04	01	4	104	041	081	19	07	07	07	11	02	21	19	094
163	3	5	18	2	07	04	02	1	157	079	080	34	09	07	12	20	07	13	34	099
164	2	5	17	2	07	05	02	1	198	102	095	45	13	11	06	46	10	15	03	108
165	3	5	17	2	06	05	02	4	093	061	088	27	05	07	12	16	11	07	33	092
166	2	5	18	2	06	05	02	4	125	057	093	25	06	09	09	25	07	19	36	077
167	2	5	18	2	36	06	01	4	129	071	079	37	08	09	08	12	08	21	21	105
168	2	5	17	2	27	05	01	5	099	064	088	18	08	02	08	16	09	04	40	095
169	2	5	18	2	24	04	07	1	206	105	105	46	16	11	11	34	07	18	09	107
170	2	5	17	2	27	04	05	5	125	059	091	25	10	07	07	18	04	04	40	085
171	2	5	18	2	24	05	08	5	133	077	094	39	10	09	10	29	11	19	13	111
172	2	5	17	2	06	01	02	1	091	045	067	14	06	07	07	17	07	04	31	089
173	2	5	18	2	32	04	09	1	136	059	081	38	08	10	06	26	09	14	27	107
174	3	5	18	2	27	04	09	4	138	082	107	38	06	06	09	33	10	09	37	097
175	2	5	18	2	32	05	01	4	081	042	080	23	05	06	06	10	07	10	29	097
176	2	5	18	2	24	04	07	3	089	032	074	17	03	04	04	06	08	09	26	085
177	2	5	18	2	36	05	03	3	109	045	088	26	06	04	04	12	07	11	33	102
178	2	5	17	2	27	08	01	4	099	071	090	33	10	10	10	17	08	04	40	083
179	2	5	17	2	27	05	01	5	101	058	090	30	04	03	07	18	10	05	39	101
180	2	5	18	2	32	05	01	5	109	071	093	37	11	13	10	22	11	02	29	090
181	3	5	15	2	03	05	05	1	132	068	094	29	09	06	07	25	07	21	24	097
182	4	5	17	2	07	08	09	1	106	062	085	22	07	07	08	18	05	15	37	098
183	2	5	17	2	23	04	05	2	163	092	099	39	13	14	11	24	08	11	07	106
184	3	5	17	2	07	04	08	1	159	082	094	29	12	12	10	20	06	23	23	118
185	4	5	17	2	07	05	07	1	153	080	093	44	09	09	12	19	05	09	04	114
186	4	5	17	2	07	04	07	1	128	083	104	43	15	10	12	27	10	08	27	101
187	4	5	17	2	36	04	09	4	063	043	084	27	02	04	06	11	06	08	36	091
188	2	5	17	2	27	05	08	4	117	074	086	41	11	09	07	27	09	03	37	104
189	3	5	16	2	27	03	08	5	130	079	109	40	16	08	12	31	05	04	29	097
190	2	5	17	2	06	04	02	1	136	079	072	33	07	08	13	21	05	12	14	096
191	2	5	17	2	27	04	08	4	117	066	102	30	09	08	09	20	05	13	23	087
192	4	5	18	2	06	03	02	1	141	075	089	28	09	08	07	19	04	23	27	101

ID	1	2	3	4	5	6	7	8	9	10	11	12	13	14	15	16	17	18	19	20
193	4	5	18	2	27	05	05	1	106	060	090	28	02	06	05	20	10	05	37	087
194	2	5	18	2	06	04	08	1	112	048	087	41	07	06	08	31	03	18	26	094
195	2	5	18	2	27	06	01	4	108	056	091	24	08	07	11	16	05	04	37	085
196	3	5	18	2	07	05	02	2	107	053	088	19	09	05	06	19	09	11	35	082
197	3	5	17	2	36	06	07	2	141	081	097	43	13	12	12	21	07	21	24	104
198	3	5	17	2	18	05	05	1	134	075	080	27	07	06	08	25	10	05	04	107
199	3	5	17	2	36	04	07	1	200	112	088	42	17	18	11	36	09	21	33	116
200	3	5	18	2	01	04	08	4	089	045	083	16	07	02	05	13	02	06	33	086
201	4	5	16	2	06	04	08	1	126	067	100	31	14	08	13	25	11	09	23	112
202	2	5	17	2	18	05	07	2	202	105	103	46	16	17	14	42	08	24	13	112
203	2	3	20	2	06	05	01	5	089	040	058	14	02	08	03	20	06	07	24	070
204	3	5	18	2	07	03	07	3	125	077	081	34	08	09	10	16	10	24	34	095
205	2	5	19	2	36	06	07	4	129	076	092	41	11	03	09	21	11	04	24	095
206	2	5	18	2	06	04	05	5	130	081	090	19	04	09	03	17	10	15	16	112
207	3	5	17	2	27	04	05	2	126	081	098	40	12	09	11	18	07	01	30	089
208	1	5	18	2	01	05	09	5	119	062	090	23	05	06	03	11	08	00	21	093
209	3	5	17	2	24	04	09	3	127	055	078	36	11	12	07	22	10	07	29	101
210	2	5	17	2	19	05	09	4	116	066	083	23	04	04	05	13	11	13	23	085
211	3	6	17	2	32	05	09	3	092	053	075	29	07	05	06	16	07	01	26	106
212	2	6	17	2	16	06	07	1	181	088	093	41	13	15	09	26	05	33	07	095
213	4	6	17	2	27	04	07	3	151	094	089	34	11	11	07	23	10	08	29	113
214	4	3	17	2	15	06	01	5	103	056	073	20	03	10	07	14	07	16	34	105
215	4	3	18	2	26	05	01	5	116	066	084	37	13	11	08	26	08	10	21	098
216	2	6	18	2	32	06	02	4	165	084	107	44	13	13	13	43	10	24	37	087
217	2	6	17	2	27	06	08	5	103	059	085	37	12	12	13	20	09	01	36	095
218	2	6	18	2	16	05	07	1	157	089	094	37	10	10	11	39	10	11	10	113
219	2	6	18	2	24	06	09	4	130	072	080	29	13	07	06	18	11	08	06	081
220	4	6	17	2	07	06	02	1	134	071	091	28	06	05	10	24	10	24	23	099
221	4	6	18	2	27	02	08	4	147	089	091	37	07	06	10	22	07	00	26	095
222	4	6	17	2	06	05	04	2	171	091	086	38	06	10	10	24	11	18	11	108
223	4	6	17	2	06	04	04	1	166	082	107	43	11	10	14	34	09	14	29	110
224	2	6	20	2	24	06	08	1	100	053	083	22	01	01	08	14	07	10	23	109
225	4	6	18	2	27	06	05	1	154	082	102	44	11	08	07	23	07	13	24	091
226	4	6	17	2	36	03	01	1	159	090	094	45	10	10	10	41	10	23	39	106
227	4	6	17	2	27	04	01	5	156	088	089	43	17	07	10	29	12	15	39	101
228	4	6	17	2	27	04	01	4	092	054	086	27	02	04	11	17	07	11	31	118
229	3	6	16	2	02	03	05	1	198	099	100	48	11	16	12	46	10	13	13	114
230	3	6	18	2	32	03	01	5	114	069	096	33	09	07	09	25	08	08	33	105
231	3	6	17	2	27	04	09	4	107	057	093	23	10	05	07	15	10	04	37	092
232	3	6	18	2	32	04	07	1	152	076	095	34	02	13	11	18	07	21	24	103
233	3	6	18	2	28	04	07	1	154	090	098	39	12	09	09	31	09	13	04	116
234	3	6	18	2	36	04	07	1	161	102	105	41	15	11	12	36	07	00	36	099
235	4	7	17	2	24	03	01	4	116	062	093	36	13	13	12	24	02	13	30	093
236	4	7	17	2	27	03	02	4	135	075	088	37	11	09	12	17	01	08	31	093
237	2	7	17	2	36	05	01	5	152	077	099	41	07	09	11	25	02	19	09	087
238	2	7	18	2	19	07	01	5	128	069	070	18	09	13	08	26	05	16	23	101
239	2	7	17	2	32	05	05	1	147	077	080	34	16	06	03	19	10	04	33	104
240	3	7	18	2	29	05	08	5	136	075	078	34	12	08	10	17	10	21	36	103

ID	1	2	3	4	5	6	7	8	9	10	11	12	13	14	15	16	17	18	19	20
241	3	8	18	2	06	05	02	3	133	065	084	29	11	12	11	19	11	06	19	103
242	2	8	17	2	27	05	09	3	121	076	090	30	10	07	10	10	07	18	29	096
243	3	8	18	2	07	04	07	2	172	107	099	46	16	08	12	18	05	07	10	100
244	3	8	17	2	36	03	01	5	126	069	090	23	05	11	09	11	10	07	34	101
245	2	8	17	2	27	07	01	5	161	082	098	45	15	13	11	24	10	02	31	091
246	3	8	17	2	27	05	01	5	124	079	085	32	11	08	10	19	09	02	33	108
247	3	8	17	2	27	04	01	5	146	073	088	39	11	07	07	15	11	03	30	103
248	3	8	18	2	01	04	01	5	164	088	097	37	11	10	11	24	03	16	34	099
249	3	8	17	2	29	04	01	5	127	074	095	38	08	10	09	20	07	06	24	105
250	2	8	17	2	24	04	09	4	117	081	095	28	13	06	10	17	06	07	34	088
251	3	8	17	2	02	04	05	1	199	098	104	44	16	18	14	34	05	22	17	105
252	3	8	18	2	18	04	05	3	158	073	090	40	13	16	12	24	07	11	19	089
253	4	8	16	2	36	05	09	2	123	080	076	39	12	11	11	22	09	20	07	098
254	4	8	17	2	32	05	01	5	117	070	079	30	06	11	12	17	11	25	40	105
255	3	8	17	2	18	05	06	1	085	059	085	26	07	07	12	05	17	23	092
256	3	8	17	2	36	07	06	1	172	089	081	44	12	10	12	25	05	19	20	098
257	4	8	17	2	18	04	05	3	182	109	096	30	14	08	06	22	03	11	06	107
258	4	8	17	2	07	04	07	2	161	084	092	32	05	09	10	21	09	15	26	109
259	4	8	17	2	26	05	07	5	157	090	099	40	13	11	09	23	12	25	26	091
260	4	8	18	2	10	03	07	1	143	080	088	40	07	10	11	31	01	20	23	100
261	3	8	16	2	32	04	08	1	153	100	098	43	16	12	12	26	10	09	37	098
262	3	8	17	2	27	04	01	5	105	065	084	27	01	08	08	08	01	01	29	092
263	4	8	16	2	29	05	08	3	066	051	082	23	04	05	07	08	05	07	34	096
264	4	8	17	2	23	07	07	4	170	094	097	43	15	15	11	39	10	31	39	116
265	4	8	17	2	06	04	05	1	189	101	100	48	15	15	13	27	08	19	11	116
266	3	9	17	2	07	03	02	1	164	078	096	41	11	11	10	23	01	20	27	090
267	3	9	18	2	27	04	09	4	072	037	072	20	08	07	09	17	07	16	26	076
268	3	5	17	2	36	05	07	5	096	045	063	21	07	06	07	05	08	21	11	106
269	4	7	17	2	07	04	02	1	120	079	092	41	11	13	11	18	07	11	11	109
270	4	7	18	2	27	05	08	5	148	075	087	22	05	06	08	18	08	12	37	110
271	4	7	17	2	07	05	07	1	167	090	105	45	14	11	11	26	01	11	27	114

APPENDIX C

Answers to selected exercises

Answers are provided here to questions involving relatively short answers. These questions were provided to aid student study and we feel that your study is further facilitated if you can easily check your work. Those questions that are more of a discussion nature are left for classroom discussion. You are also encouraged to write up those exercises involving the computer even if write-ups are not required by the instructor. These computer exercises form an integral part of this course and experience has shown that maximum benefit is derived from the computer work if you systematically consider the purpose, results, and conclusions of the Monte Carlo experiments and other computer exercises.

Chapter Two

1. Nominal   Ordinal   Interval   Ratio
    2, 4, 5, 7   1, 8   6 (as coded), 9–20   3, 6 (as actual pounds)

2. 17

3. (a) Variable 5 is not measured on an interval or ratio scale as required to employ the mean.
   (b) Utilize the mode, or still better, the entire distribution for the nominal scale, possibly converted to proportions.

4.

Score	02	03	04	05	06	07	08	09	10	11	12
Frequency	1	1	2	3	6	8	6	4	10	4	3

5. *Sense.* The sample upon which the means are based is random and is expected, therefore, to approximate the actual means of the entire population.
   Nonsense. It is entirely possible that the population sampled is quite heterogeneous and the means, therefore, do not present a complete picture of the typical student.

Chapter Three

2. This illustrates the fact that the sum of the deviations from the mean is zero.

$$M_x = 7.9 \qquad x = \begin{array}{r} 2.1 \\ -3.9 \\ 1.1 \\ -3.9 \\ 3.1 \\ -1.9 \\ 3.1 \\ 1.1 \\ -1.9 \\ 1.1 \\ \hline \end{array}$$
$$\sum x = 0$$

3.

X	f	$f \cdot X$	$(f \cdot X) \cdot X$
1	4	4	4
2	2	4	8
3	2	6	18
4	2	8	32
5	10	50	250
	$N = 20$	$\sum X = 72$	$\sum X^2 = 312$

$$s_x^2 = \frac{1}{20}\left[312 - \frac{(72)^2}{20}\right]$$
$$= \frac{1}{20}(312 - 259.2)$$
$$= 2.64$$

4. $\sqrt{2.64} = 1.6$ Tchebysheff's theorem   Let $k = 1.5$
$M_x = 3.6$

$M_x \pm k\sigma_x$ encloses at least $\dfrac{1}{1 - k^2}$ part of N scores

$3.6 \pm 1.5(1.6)$ encloses at least $1 - \dfrac{1}{2.25}$ part of 20 scores

(1.2 to 6.0)   encloses at least $\frac{5}{9}$ part of 20 scores

This certainly holds.

5. $T = 10z + 50$
$= 70$

$z = \dfrac{x}{s} = \dfrac{175 - 159}{8} = \dfrac{16}{8} = 2$

6. $1 - \dfrac{1}{(1.2)^2} = \dfrac{11}{36}$

7. $z = 1.2$
$T = 10(1.2) + 50 = 62$

ANSWERS TO SELECTED EXERCISES

8. One group may be heterogeneous (more variable) with respect to readiness while the other may be homogeneous with respect to the same variable. Thus one plan may be insufficient for both groups.

Chapter Four

1. J = 1
   SUM = 19.5
   X(1) = 3.2
   SUM = 19.5 + 3.2
   SUM = 22.7

2. A = 38

3. K3 = 4
   M2 = 11 * 4 = 44
   J = 44/20 = 2
   J = 2

4. $R = \dfrac{A + BX}{C + DX}$

5. M = 5
   K = 1

   Therefore computer will go to statement number 7

6. (a) Invalid—Mixed mode
   (b) Invalid—Mixed mode
   (c) Valid—An exponent can be an integer constant.

7. Read (5, 7) ID, LWT, JSOC
   Format (1X, I3, I4X, I2, 33X, I2)

8. Read (5, 8) ID, WT, SOC
   Format (I4, 14X, F2.0, 33X, F2.0)

Chapter Five

1. Before writing output, change

   $$Y = \text{Random (0)}$$

   to

   $$IY = 100* \text{ Random (0)}$$

   (This declares Y an integer by altering its form to IY)

2.

X	f	fX	(fX)X
6	1	6	36
5	1	5	25
4	1	4	16
33	1	3	9
2	1	2	4
1	1	1	1
N = 6		$\sum X = 21$	$\sum X^2 = 91$

$\sigma_x^2 = \dfrac{1}{6}\left[91 - \dfrac{(21)^2}{6}\right]$
$= \dfrac{1}{6}(91 - 73.5)$
$= 2.92$

3. Coin

5. The *expected value* and the *parameter* of a distribution are in a sense synonymous. The expected value is the theoretical statistic known as the parameter.

6. A sampling distribution is that which the scientist actually observes when he takes a sample of the total population, while the theoretical distribution is that which the laws of probability state will actually occur.

8. It illustrated the bias that occurs in the sampling distribution of the sample variance and the ability to correct that error by employing the unbiased estimate of the variance.

Chapter Six

1. $p(X) = \binom{5}{5} p^5 q^0$

   $= \dfrac{5!}{5!0!} \left(\dfrac{1}{4}\right)^5 = \dfrac{1}{1024} = 0.0010$

   Where X = event that 5 hearts drawn in 5 draws

2. $p(X)$. The sum of all of the $p(X)$ must be one. The fact that when the variable is continuous the $p(X)$ are infinite in number make $p(X)$ useless with continuous variables.

3. $p(X) = \dfrac{16}{271}$ where X = response is answer number 3

4. Median of group is 81

   $\bar{X} = 79.077$ X = the event that boy scores below grade-sex mean on Information Test

   $P(X) = .487$

   $P(X) = .50$ if the mean and median of the distribution are the same

5. $\binom{8}{4} = \dfrac{8!}{4!4!} = \dfrac{8 \cdot 7 \cdot 6 \cdot 5 \cdot 4 \cdot 3 \cdot 2}{4 \cdot 3 \cdot 2 \cdot 4 \cdot 3 \cdot 2} = 70$

6. Inverse cumulative proportion implies the proportion of scores that are "greater" than or equal to any given score in the distribution.

7. Given a uniform distribution, the sum of the probabilities of the occurence $X = j$ (where $j = 1$ through $j = k$) is 1.

ANSWERS TO SELECTED EXERCISES

Chapter Seven

2. (1, 1) (2, 1) . . . . . (6, 1) X = sum of dice
    (1, 2) (2, 2) . . . . . (6, 2)
    (1, 3) . . . .
    . . . .
    (1, 6) (2, 6) . . . . . (6, 6)

X	f
2	1
3	2
4	3
5	4
6	5
7	6
8	5
9	4
10	3
11	2
12	1
	36

3. $X = 42$
    $N = 52$
    $Z = [42 - .5(52) + .5]/\sqrt{.25(52)}$
    $= 16.5/\sqrt{13}$
    $= 4.58$
    $P(Z \leq 4.58) > .99$

5. $N = 234$ Mean and SD required-use program
    $X = 141$
    $$Z = \frac{141 - 156.504}{35.155} = -.443$$
    $P(Z \leq -.4) = .345$

6. Binomial is a discrete distribution while the normal is continuous.

7. $P(Z = -1.8) = .036$
    $P(Z = 1.2) = .885$
    $.885 - .036 = .849$
    Therefore about 85 percent will fall between -1.8 and 1.2.

Chapter Eight

1. (a) 841
    (b) 532
    (c) Boy (note the standard deviation for boys)
    (d) Yes

2. 15.9 percent
    Yes

3.

	Boys	Girls
Mean	10.410	9.380
Standard Deviation	3.886	3.840
N	234	271

$$S_{M_B - M_G} = \sqrt{.0645 + .0544}$$
$$= .345$$

$$S_{M_B}^2 = \frac{(3.886)^2}{234} = \frac{15.101}{234} = .0645$$

$$S_{M_G}^2 = \frac{(3.840)^2}{271} = \frac{14.746}{271} = .0544$$

95 percent confidence limits:
1.03 ± 2(.345)

The difference between TALENT Creativity means for males and females lies in the interval .340 to 1.72 with 95 percent probability. The best estimate of difference in means is 1.03. The hypothesis that $\mu_1 = \mu_2$ can be rejected with 95 percent confidence, since 0.0 does not lie within the obtained interval.

5. $\sigma_M^2 = \dfrac{10}{40} = .25 \quad \sigma_M = .50$

$$Z = \frac{51 - 50}{.5} = 2.00$$

Therefore the probability of drawing a sample mean equal to or greater than 51 is .0228.

6. The central-limit theorem assures us that sample means of X will be normally distributed even if the variable X is not normally distributed.

Chapter Ten

1. $\sigma^2 = 9.0$

$$\chi_{233}^2 = \frac{233(7.586)^2}{9} = \frac{13,409.15}{9} = 1489.9$$

$S^2 = 7.586$

$N = 234$

for $\chi_{240}^2 = 293, P = .99$

Therefore, we reject the null hypothesis since our sampling procedure is suspect.

ANSWERS TO SELECTED EXERCISES

2.

$N = 505$

	f Males		f Females	
	O	E	O	E
1	70	(70.0)	81	(81.0)
2	33	(25.9)	23	(30.1)
3	34	(23.2)	16	(26.8)
4	44	(44.9)	53	(52.1)
5	53	(70.0)	98	(81.0)
Total	234	234	271	271

	Males $(O-E)^2/E$	Females $(O-E)^2/E$
1	0	0
2	1.95	1.67
3	5.03	4.35
4	.02	.02
5	4.13	3.57

$\chi_4^2 = 20.74$
$P > .999$
$\therefore$ Reject H_0

Chapter Eleven

1. .80

2. Mean 65 45
   N   16 16

$t = \dfrac{65 - 45}{10}$, n.d.f. = 30

$t_{30} = 2$

$P(t_{30}) > .95$

$S_{M_A - M_B} = 10$

Method A is probably more effective than method B. There are less than five chances in 100 that the differences are attributable to chance.

4. $S_{M_1 - M_2} = \sqrt{\left(\dfrac{\sum x_1^2 + \sum x_2^2}{N_1 + N_2 - 2}\right)\left(\dfrac{1}{N_1} + \dfrac{1}{N_2}\right)}$ If $N_1 = N_2 = N$ then

$= \sqrt{\dfrac{\sum x_1^2 + \sum x_2^2}{2N - 2} \cdot \dfrac{2}{N}}$

$= \sqrt{\dfrac{\sum x_1^2 + \sum x_2^2}{N(N-1)}} = \sqrt{\dfrac{\sum(X - M_1)^2}{N(N-1)} + \dfrac{\sum(X - M_2)^2}{N(N-1)}}$

$= \sqrt{\dfrac{S_{X_1}^2}{N} + \dfrac{S_{X_2}^2}{N}} = \sqrt{S_{M_1}^2 + S_{M_2}^2}$

5. 82 out of 200 times the *t* was significant at the .05 level. Therefore, 118 out of 200 it was not significant at the .05 level. If we reject the null hypothesis of no difference, we shall be correct 41 percent of the time thus implying that 59 percent of the time we shall err, thus making a Type II error. With XMD = 2.

   172 out of 200 times the *t* was significant at .05 level.

6. Where P represents the probability of getting a *t* equal to or smaller than the obtained *t*, α represents the probability of getting a *t* as large or larger than the obtained *t*. $\alpha = 1 - P$

Chapter Thirteen

1. X = Reading Comprehension, Y = Socioeconomic Status
   $X = .284\ Y + 5.797$
   Illustrate the male-female regression differences graphically by plotting the two regression.

2. Determine the mean of reading comprehension for each sex, and test the significance of difference between the resulting means.

Chapter Fourteen

1. Use Z since $(.55)^2 = .3025$ as opposed to $(-.72)^2 = .5184$. Therefore, we can predict 52 percent of the unknown variation in X using Z.

2. Negatively. If it is negative, this implies that the trend is toward the older twelfth-grade students having lower reading comprehension. It is negative!

3. NO!!

APPENDIX D

Summary of mathematical concepts necessary for understanding elementary statistical concepts[1]

The purpose of this appendix is to give students some idea of the mathematical background essential to understanding elementary statistics. It is not designed as a teaching device but rather as a device to help the student evaluate his own mathematical background and determine whether or not further study of these topics seems necessary. A reference to a good text for further study is suggested at the end.

1. Addition, subtraction, multiplication and division of signed numbers

Examples:

(a) $(-5) + (-2) = -7$
(b) $(-5) - (-2) = -3$
(c) $(-5) + (2) = -3$
(d) $(-5)(-2) = 10$
(e) $(-5)(2) = -10$
(f) $(5)(-2) = -10$
(g) $(-10)/(-2) = 5$
(h) $(-10)/(2) = -5$

2. Exponentiation

Examples:

(a) $2^2 = 2 \times 2 = 4$
(b) $3^3 = 3 \times 3 \times 3 = 27$
(c) $4^{1/2} = \sqrt{4} = 2$
(d) $1^0 = 2^0 = 3^0 = 1$ (Any number with an exponent of zero is equal to 1)
(e) $(2^2)^3 = 2^6$
(f) $(3^4) \times (3^5) = 3^9$

---

* We are indebted to Mr. James E. Carlson of the University of Pittsburgh for providing this appendix.

267

3. **Symbolic representation of numbers by letters**

   Examples:

   If
   $$X = 2$$
   $$Y = -3$$
   $$A = X + Y$$
   Then
   $$A = 2 + (-3) = -1$$

4. **Inequalities**

   Examples:

   $X < Y$ (X less than Y)
   $X \leq Y$ (X less than or equal to Y)
   $X \neq Y$ (X not equal to Y)

5. **Proportions**

   Examples:

   The proportion of hearts in a deck of cards is

   $$\frac{13}{52} = .25$$

6. **Use of a symbol to represent a set of numbers; subscripts**

   The set of scores of 10 people on a particular test can be called X. Particular scores can be represented by the letter X with a subscript.

   Examples:

   X is the set of scores   5, 7, 9, 6, 4, 8, 7, 5, 6, 4
   $X_1 = 5$, $X_2 = 7$, $X_3 = 9$,   etc.

   In general:

   X_i (where i can be any integer from 1 to 10) can be used to denote a particular person's score from the above set. For example, the eighth person's score ($i = 8$) is

   $$X_8 = 5$$

   X_i ($i = 1, N$) is a symbolic representation of a set of N different values of the variable X.

7. **Equations expressing relationships between variables**

   Examples:

   $Y = aX + b$ where X and Y are variables, a and b are constants

   Given the equation relating two variables and the values of the constants, we can

SUMMARY OF MATHEMATICAL CONCEPTS

determine the value of one of the variables for any given value of the other. For example,

$$Y = 2X + 3$$

When
$$X = 0,$$
$$Y = 2(0) + 3 = 3$$

When
$$X = -1,$$
$$Y = 2(-1) + 3 = 1$$

When
$$X = 1,$$
$$Y = 2(1) + 3 = 5$$

The following pairs of values can thus be determined

X	Y
-3	-3
-2	-1
-1	1
0	3
1	5
2	7

These pairs of values of the variables can also be plotted as points on a graph. Thus $Y = 2X + 3$ is the equation of the line plotted above. Note that the constant b is the Y-intercept, that is, the value of Y where the line crosses the Y axis. The coefficient a indicates the degree of inclination of the line. When a is 2, an increase of 1 unit in X is associated with a change of 2 units in Y.

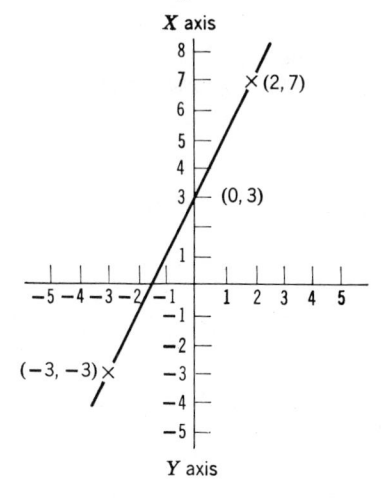

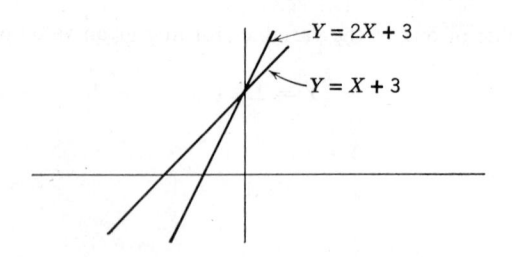

Changing only the coefficient a in an equation thus changes only the inclination of the line and not the Y-intercept.

Changing only the intercept b in an equation thus changes only the Y intercept of the line and not the degree of inclination.

If the exponents of both variables are 1, the plot will result on a straight line. Thus such equations as:

$$Y = 2X + 3$$

and

$$Z = \frac{W}{5} + 10$$

are known as linear equations.

If either variable has an exponent other than 1, the plot will result in a curved line. Thus the following are not linear equations.

$$Y = 2X^2 + 6$$
$$W = 2Z^3 + 3Z^2 + 5$$
$$\frac{A^2}{4} = \frac{B^2}{9}$$

8. Summation

The Greek letter Σ (capital sigma) is used as an operator designating "the sum of." The range of values summed can be indicated.

Examples:

(a) $\sum_{i=1}^{N} Z_i$ All values of the variable Z from Z_1 to Z_N are added together

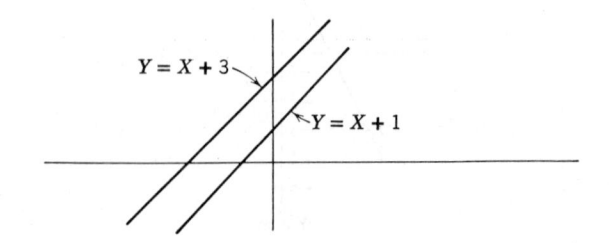

SUMMARY OF MATHEMATICAL CONCEPTS

(b) Using the set of numbers in 6 above, where $N = 10$.

$$\sum_{i=1}^{N} X_i = X_1 + X_2 + X_3 + X_4 + X_5 + X_6 + X_7 + X_8 + X_9 + X_{10}$$
$$= 5 + 7 + 9 + 6 + 4 + 8 + 7 + 5 + 6 + 4$$
$$= 61$$

Note. Where it is obvious that all values of a variable are to be summed the notation may be abbreviated to

$$\sum_i X_i \quad \text{or} \quad \sum X$$

9. Summation rules

(a) Distribution of a summation sign

$$\sum_{i=1}^{N} (X_i + Y_i) = \sum_{i=1}^{N} X_i + \sum_{i=1}^{N} Y_i$$

Or in the abbreviated notation

$$\sum (X + Y) = \sum X + \sum Y$$

Example:

If X is the set
$$5, 7, 9, 6, 4$$
and Y is the set
$$12, 10, 15, 16, 11$$
then
$$\sum (X + Y) = (5 + 12) + (7 + 10) + (9 + 15) + (6 + 16) + (4 + 11) = 95$$
$$\sum X + \sum Y = (5 + 7 + 9 + 6 + 4) + (12 + 10 + 15 + 16 + 11) = 95$$

(b) Summing over a constant

$$\sum_{i=1}^{N} C = NC$$

Example:

$$C = 3, \quad N = 4$$
$$\sum C = 3 + 3 + 3 + 3 = 12$$
$$NC = 4 \times 3 = 12$$

(c) Summing a constant multiplier of a variable

$$\sum_{i=1}^{N} CX_i = C \sum_{i=1}^{N} X_i$$

Examples:

$C = 3$, X is the set 10, 9, 12
$$\sum CX = (3 \times 10) + (3 \times 9) + (3 \times 12) = 93$$
$$C \sum X = 3(10 + 9 + 12) = 93$$

(d) Combination of summation rules

$$\sum_{i=1}^{N}(C + KX_i) = \sum_{i=1}^{N} C + \sum_{i=1}^{N} KX_i = NC + K\sum_{i=1}^{N} X_i$$

Examples:

$C = 4$; X is the set 5, 8, 6; $K = 2$

$\sum (C + KX) = (4 + 2 \times 5) + (4 + 2 \times 8) + (4 + 2 \times 6) = 50$

$NC + K\sum X = 3 \times 4 + 2(5 + 8 + 6) = 50$

10. **Algebraic manipulation of symbols**
Examples:
(a) $A + B + C = (A + B) + C = A + (B + C) = C + A + B$, etc.
(b) $A \times B \times C = (A \times B) \times C = A \times (B \times C) = B \times (C \times A)$, etc.
(c) $AB + AC = A(B + C)$
(d) $\dfrac{A + B}{C} = \dfrac{A}{C} + \dfrac{B}{C}$
(e) $\dfrac{A}{B/C} = A \times \dfrac{C}{B} = \dfrac{AC}{B}$
(f) $\dfrac{A}{B} + \dfrac{C}{B^2} = \dfrac{BA}{B^2} + \dfrac{C}{B^2} = \dfrac{BA + C}{B^2}$
(g) $\left(\dfrac{A}{B}\right)^2 = \dfrac{A^2}{B^2}$

11. **Use of parentheses to denote order of algebraic operations**
Examples:
(a) $\dfrac{(A + B)}{C}$ A is first added to B, then the sum is divided by C

(b) $A(B + C)^2$ B is first added to C, next the sum is squared, then the squared sum is multiplied by A

12. **Expansion of a binomial**
(a) $(X + Y)^2 = X^2 + 2XY + Y^2$
(b) $(X - Y)^2 = X^2 - 2XY + Y^2$

13. **Combination of summation and binomial expansion**

$$\sum (X - Y)^2 = \sum (X^2 - 2XY + Y^2)$$
$$= \sum X^2 - 2\sum XY + \sum Y^2$$

SUMMARY OF MATHEMATICAL CONCEPTS

Note

$$\sum X = X_1 + X_2 + \cdots + X_N$$
$$\sum X^2 = X_1^2 + X_2^2 + \cdots + X_N^2$$
$$\sum XY = X_1 Y_1 + X_2 Y_2 + \cdots + X_N Y_N$$
$$(\sum X)^2 = (X_1 + X_2 + \cdots + X_N)^2$$

14. The student in statistics should be able to integrate the above concepts so that he can follow proofs such as the following

Given:

X is a set of N numbers

M is the average of the N numbers $\left(M = \dfrac{\sum X}{N}\right)$

Show that:

$$\frac{\sum (X - M)^2}{N - 1} = \frac{N \sum X^2 - (\sum X)^2}{N(N - 1)}$$

Proof:

$$\frac{\sum (X - M)^2}{N - 1} = \frac{\sum (X^2 - 2XM + M^2)}{N - 1}$$

$$= \frac{\sum X^2 - 2M \sum X + NM^2}{N - 1}$$

$$= \frac{\sum X^2 - 2(\sum X/N) \sum X + N(\sum X/N)^2}{N - 1}$$

$$= \frac{\sum X^2 - 2(\sum X)^2/N + N(\sum X)^2/N^2}{N - 1}$$

$$= \frac{\sum X^2 - 2(\sum X)^2/N + (\sum X)^2/N}{N - 1}$$

$$= \frac{N \sum X^2/N - 2(\sum X)^2/N + (\sum X)^2/N}{N - 1}$$

$$= \frac{N \sum X^2/N - (\sum X)^2/N}{N - 1} = \frac{N \sum X^2 - (\sum X)^2/N}{N - 1}$$

$$= \frac{N \sum X^2 - (\sum X)^2}{N(N - 1)}$$

References for further study of the above concepts

Walker, Helen M., *Mathematics Essential for Elementary Statistics*. New York: Holt, 1951.

APPENDIX E

Theoretical distributions*

Table 7.1 Cumulative Probabilities $P(z)$ of the $N(0, 1)$ Distribution[a]

z	$P(z)$	z	$P(z)$
−3.5	.00023	.1	.540
−3.0	.00135	.2	.579
−2.8	.00256	.3	.618
−2.6	.00466	.4	.655
−2.5	.00621	.5	.691
−2.4	.00820	.6	.726
−2.2	.0139	.8	.788
−2.0	.0228	1.0	.841
−1.8	.0359	1.2	.885
−1.6	.0548	1.4	.919
−1.5	.0668	1.5	.933
−1.4	.0808	1.6	.945
−1.2	.115	1.8	.964
−1.0	.159	2.0	.977
−.8	.212	2.2	.986
−.6	.274	2.4	.992
−.5	.309	2.5	.9938
−.4	.345	2.6	.9953
−.3	.382	2.8	.9974
−.2	.421	3.0	.9987
−.1	.460	3.5	.99977
0.0	.500	4.0	.99997

[a] Values obtained from Abramowitz and Stegun (1964), pp. 966–973.

These frequently used tables have been repeated here for easy reference.

Table 9.1 Cumulative Probabilities for χ^2 [a]

n.d.f. P:	.01	.05	.10	.90	.95	.99	.999
1	.0002	.004	.016	2.71	3.84	6.63	10.8
2	.020	.103	.211	4.61	5.99	9.21	13.8
3	.115	.352	.584	6.25	7.81	11.3	16.3
4	.297	.711	1.06	7.78	9.49	13.3	18.5
5	.554	1.15	1.61	9.24	11.1	15.1	20.5
6	.872	1.64	2.20	10.6	12.6	16.8	22.5
7	1.24	2.17	2.83	12.0	14.1	18.5	24.3
8	1.65	2.73	3.49	13.4	15.5	20.1	26.1
9	2.09	3.33	4.17	14.7	16.9	21.7	27.9
10	2.56	3.94	4.87	16.0	18.3	23.2	29.6
11	3.05	4.57	5.58	17.3	19.7	24.7	31.3
12	3.57	5.23	6.30	18.5	21.0	26.2	32.9
13	4.11	5.89	7.04	19.8	22.4	27.7	34.5
14	4.66	6.57	7.79	21.1	23.7	29.1	36.1
15	5.23	7.26	8.55	22.3	25.0	30.6	37.7
16	5.81	7.96	9.31	23.5	26.3	32.0	39.3
17	6.41	8.67	10.1	24.8	27.6	33.4	40.8
18	7.01	9.39	10.9	26.0	28.9	34.8	42.3
19	7.63	10.1	11.7	27.2	30.1	36.2	43.8
20	8.26	10.9	12.4	28.4	31.4	37.6	45.3
22	9.54	12.3	14.0	30.8	33.9	40.3	48.3
24	10.9	13.8	15.7	33.2	36.4	43.0	51.2
26	12.2	15.4	17.3	35.6	38.9	45.6	54.1
28	13.6	16.9	18.9	37.9	41.3	48.3	56.9
30	15.0	18.5	20.6	40.3	43.8	50.9	59.7
40	22.2	26.5	29.1	51.8	55.8	63.7	73.4
60	37.5	43.2	46.5	74.4	79.1	88.4	99.6
80	53.5	60.4	64.3	96.6	102.	112.	125.
100	70.1	77.9	82.4	118.	124.	136.	149.

[a] Values selected from Abramowitz and Stegun (1964), pp. 984–985.

Table 11.1 Cumulative Probabilities for t[a]

n.d.f. P:	.001	.01	.05	.10	.90	.95	.975	.99	.999
1	−318.	−31.8	−6.31	−3.08	3.08	6.31	12.7	31.8	318.
2	−22.3	−6.97	−2.92	−1.89	1.89	2.92	4.30	6.97	22.3
3	−10.2	−4.54	−2.35	−1.64	1.64	2.35	3.18	4.54	10.2
4	−7.17	−3.75	−2.13	−1.53	1.53	2.13	2.78	3.75	7.17
5	−5.89	−3.37	−2.02	−1.48	1.48	2.02	2.57	3.37	5.89
6	−5.21	−3.14	−1.94	−1.44	1.44	1.94	2.45	3.14	5.21
7	−4.79	−3.00	−1.90	−1.42	1.42	1.90	2.37	3.00	4.79
8	−4.50	−2.90	−1.86	−1.40	1.40	1.86	2.31	2.90	4.50
9	−4.30	−2.82	−1.83	−1.38	1.38	1.83	2.26	2.82	4.30
10	−4.14	−2.76	−1.81	−1.37	1.37	1.81	2.23	2.76	4.14
11	−4.03	−2.72	−1.80	−1.36	1.36	1.80	2.20	2.72	4.03
12	−3.93	−2.68	−1.78	−1.36	1.36	1.78	2.18	2.68	3.93
13	−3.85	−2.65	−1.77	−1.35	1.35	1.77	2.16	2.65	3.85
14	−3.79	−2.62	−1.76	−1.35	1.35	1.76	2.15	2.62	3.79
15	−3.73	−2.60	−1.75	−1.34	1.34	1.75	2.13	2.60	3.73
16	−3.69	−2.58	−1.75	−1.34	1.34	1.75	2.12	2.58	3.69
17	−3.65	−2.57	−1.74	−1.33	1.33	1.74	2.11	2.57	3.65
18	−3.61	−2.55	−1.73	−1.33	1.33	1.73	2.10	2.55	3.61
19	−3.58	−2.54	−1.73	−1.33	1.33	1.73	2.09	2.54	3.58
20	−3.55	−2.53	−1.73	−1.33	1.33	1.73	2.09	2.53	3.55
22	−3.51	−2.51	−1.72	−1.32	1.32	1.72	2.07	2.51	3.51
24	−3.47	−2.50	−1.71	−1.32	1.32	1.71	2.06	2.50	3.47
26	−3.44	−2.48	−1.71	−1.32	1.32	1.71	2.06	2.48	3.44
28	−3.41	−2.47	−1.70	−1.31	1.31	1.70	2.05	2.47	3.41
30	−3.39	−2.46	−1.70	−1.31	1.31	1.70	2.04	2.46	3.39
60	−3.23	−2.39	−1.67	−1.30	1.30	1.67	2.00	2.39	3.23
120	−3.16	−2.36	−1.66	−1.29	1.29	1.66	1.98	2.36	3.16
∞	−3.09	−2.33	−1.65	−1.28	1.28	1.65	1.96	2.33	3.09

[a] Values selected from Abramowitz and Stegun (1964), p. 990.

Index

Abramowitz, M., 113, 145, 176, 240
accumulations, 37, 38
adding, *see* summation
Ahmann, J. C., 191, 241
Alluisi, 68
alpha level, 135, 168
analysis of variance, 240
answers to exercises, 259–266
area sampling, 238

Bauer, R. A., 135, 240
Bernoulli's theorem, 104, 128
Bernoulli-type experiment, 96, 99, 104
binomial distribution, 96, 101, 104, 115
bivariate distribution, 163
bivariate normal distribution, 219, 226
Bloom, B. S., 240

case study, 170
centour, 220
central-limit theorem, 116, 118, 128, 140, 144, 146
central tendency, 27
centroid, 220
chi-square, 110, 144, 157
classification, 220
Cochran, W. G., 238, 240
coefficients of the binomial expansion, 98
Cohen, J., 173, 240
computing formulas, 37, 210, 219
confidence level, 135
confidence limits, *or* interval, 127, 137, 138, 140, 177, 214
Cooley, W. W., 7, 135, 221, 240, 241
Cooperative English Tests, 30
Cooperative Science Tests, 5
correlation, 163, 215
correlation coefficient, 217, 225, 232

correlation matrix, 224
covariance, 7
cumulative frequency distribution, 26

density function, 103, 226
descriptive statistics, 29, 40
design, 5
deviation score, 35
differential calculus, 207
digital computer, 49

Edwards, A. L., 240
ellipse, 215, 220
experiment, 2, 16, 238
extra-sensory perception (ESP), 93

factorials, 98
Feigenbaum, E. A., 50, 240
Feldman, J., 50, 240
figures
  1.1, 15
  1.2, 15
  1.3, 16
  2.1, 22
  2.2, 23
  2.3, 23
  2.4, 24
  2.5, 27
  3.1, 35
  3.2, 41
  3.3, 42
  4.1, 50
  4.2, 61
  4.3, 63
  5.1, 86
  7.1, 108
  7.2, 111
  7.3, 113

7.4, 114
7.5, 115
7.6, 115
7.7, 117
7.8, 117
7.9, 119
7.10, 119
7.11, 121
8.1, 130
8.2, 130
8.3, 132
8.4, 132
8.5, 134
9.1, 147
11.1, 175
11.2, 176
13.1, 202
13.2, 203
13.3, 205
13.4, 211
13.5, 212
14.1, 216
14.2, 217
14.3, 218
14.4, 220
14.5, 220
14.6, 221
Fisher, R. A., 186, 230, 240
Fisherian statistics, 174
Flanagan, J. C., 7, 11, 240, 241
FORTRAN, 52
frequency distribution, 21, 31
frequency polygon, 23

Galler, B. A., 80, 241
goodness of fit, 144, 149, 157
Graybill, F. A., 117, 234, 241
Green, B. F., 51, 80, 241
Gribbons, W. D., 20
grouped data frequency distribution, 24

Hays, W. L., 234, 241
histogram, 23
Hollerith field, 60
homoscedasticity, 210
Huff, D., 23, 241
hypothesis testing, 234

IBM, 67, 81, 241
interval estimation, 234

Jones, K. J., 135, 240

Kempthorne, D., 186, 241
Kerlinger, F. N., 17, 241
Kinsey studies, 239
Klopfer, L, E., 4, 240
kurtosis, 123

law of larger numbers, 118, 128
least squares, 39
linearity of array means, 204
linear transformation, 44
Lohnes, P. R., 11, 30, 45, 46, 221, 240, 241

McIntire, P. H., 30, 45, 46, 241
marginal distribution, 165
mathematics review, 267–273
maximum likelihood, 233
mean, 28, 29, 68, 132, 231
mean differences, 33, 138, 177
measurement, 18
median, 27
mode, 27
modulo 1, 81
Monte Carlo method, 87, 89, 153, 179, 223
Mood, A. M., 117, 234, 241
multivariate analysis, 240

Neidt, C. O., 191, 241
nonparametric statistics, 187, 194
normal distribution, 107, 110, 112, 146, 177, 219
norms, 30
null hypothesis, 100, 134, 164, 168, 237
number of degrees of freedom, 145, 149, 167, 177

ogive, 26, 121
one-tailed test, 136, 170
Otis Mental Ability Test, 5

parameter, 29, 126, 232
Parzen, E., 87
Pascal's triangle, 97
Pearson, K., 149
point estimation, 234
population, 28, 231
power, 170, 180
power-residue method, 81, 89
prediction, 198

INDEX

probability, 25, 99, 102, 134, 186
Project HOSC, 4, 166, 238
Project TALENT, 7, 29, 83, 213, 225, 238, 245–258
proportion, 162
pseudo-random fraction, 81
punched card, 14, 61

quantification, 231

RAND Corporation, 90
random assignment, 2, 7, 11, 16, 239
random behavior, 93
randomization, 169, 186, 194, 232
randomization tests, 186
random normal deviate, 121, 145, 154
random number generation, 79
random sampling, 12, 16, 29, 238
random variable, 101, 127, 146, 168
  continuous, 102
  discrete, 102
Readiness for Vocational Planning (RVP), 20
regression, 204, 212, 213
regression equation, 213, 214
regression slope, 213, 215, 225
relationship, 215, 236
reliability, 5
research style, 156

sample, 29, 230
sampling distribution, 79, 127, 132, 144, 223
sampling experiment, 79
scales, 19, 31
  interval, 20, 27
  nominal, 19
  ordinal, 19
  ranking, 20
  ratio, 21, 27
School and College Ability Tests (SCAT), 30
Shaycoft, M. F., 11, 241
skewness, 122, 145
Siegel, S., 187, 241
slope, 206
small-sample statistics, 174
standard deviation, 39, 40, 47
standard error, 126, 128, 139, 178, 204, 234
standard error of estimate, 209, 213, 214, 225, 234
standard scores, 43, 48, 113, 225
statistic, 3, 28, 230, 231
statistical inference, 126, 135, 136, 169, 214, 233
Stegun, I. A., 113, 145, 176, 240
Stouffer, S., 239
stratified random sample, 165, 238
"Student," 178
"Studentized" statistics, 174
summation, 28, 36
sum of squares, 38, 39
survey, 2, 7, 16, 239
symbol conventions, 243

tables
1.1, 6
1.2, 9–10
1.3, 13
1.4, 13
2.1, 22
2.2, 24
2.3, 25
2.4, 25
2.5, 26
2.6, 30
2.7, 30
2.8, 31
3.1, 37
3.2, 45
3.3, 46
5.1, 80
5.2, 88
5.3, 88
6.1, 94
6.2, 94
6.3, 97
7.1, 112
7.2, 118
7.3, 118
7.4, 122
8.1, 129
8.2, 131
8.3, 131
9.1, 146
9.2, 147
9.3, 148
9.4, 148
9.5, 148
9.6, 150
9.7, 151

9.8, 152
9.9, 152
9.10, 154
9.11, 154
9.12, 155
9.13, 156
10.1, 164
10.2, 166
10.3, 166
10.4, 167
10.5, 171
11.1, 176
11.2, 179
11.3, 180
11.4, 181
12.1, 187
12.2, 189
12.3, 192–194
13.1, 198
13.2, 200
13.3, 200
13.4, 201
13.6, 205
13.7, 207
14.1, 223
14.2, 224
14.3, 225
theoretical z, x^2, and t distributions, 274
Tchebysheff's theorem, 40, 44
t distribution, 174, 224

Terman, L., 33, 241
Test on Understanding Science (TOUS), 5, 20
theoretical distributions, 100, 101, 105 144
theory building, 235
Travers, R. M. W., 238, 241
tree structure, 201
T score, 45, 48
Turing's theorem, 50
two-tailed test, 136, 145
type I error, 168, 170, 172
type II error, 170, 171, 172, 180

uncertainty, 232
uniform distribution, 89, 104, 131
unimodality, 27
United States Bureau of Standards, 112
United States Census Bureau, 239

validity, 5
variable, 21, 102
variance, 35, 36, 46, 47, 70, 87, 161, 233
variance error, 128

Wert, J. E., 191, 241
Wingersky, B., 83

z score, 43, 48, 113